MODULAR REPRESENTATION THEORY OF FINITE GROUPS

MODULAR REPRESENTATION THEORY OF FINITE GROUPS

Editor

Alex Wilson

scitus academics

Modular Representation Theory of Finite Groups
Edited by **Alex Wilson**

ISBN: 978-1-68117-646-8
Library of Congress Control Number: 2016946508

SCITUS Academics LLC,
616, Corporate Way, Suite 2, 4766,
Valley Cottage, NY 10989

www.scitusacademics.com

Notice

Printed in United States of America on Acid Free Paper

PREFACE

In mathematics, representation theory is a technique for analyzing abstract groups in terms of groups of linear transformations. Representation theory studies maps from groups into the general linear group of a finite-dimensional vector space. For finite groups the theory comes in two distinct flavors. In the 'semisimple case' (for example over the field of complex numbers) one can use character theory to completely understand the representations. This by far is not sufficient when the characteristic of the field divides the order of the group. Modular representation theory is a branch of mathematics, and that part of representation theory that studies linear representations of finite groups over a field K of positive characteristic. As well as having applications to group theory, modular representations arise naturally in other branches of mathematics, such as algebraic geometry, coding theory, combinatorics and number theory.

This book aims to the modular representation theory of finite groups from an algebraic point of view, regarding representations as modules over the group algebra.

TABLE OF CONTENTS

1

SOME PROBLEMS IN COMPUTATIONAL REPRESENTATION THEORY

Gerhard O. Michler

Department of Mathematics, University of Essen, 4300 Essen 1, Federal Republic of Germany

1. INTRODUCTION

Construction of the irreducible representations of a finite group G requires a detailed knowledge of the group structure of G. For the study of concrete examples it is often extremely useful to apply the algorithms contained in the general computer algebra systems. So far, the systems CAYLEY of Cannon (1984) and CAS of Neub/iser, Pahlings & Plesken (1984) have played an important role in the study and application of special group representations.

In addition, the computer may also be used for the computation of the generic character tables of some finite groups of Lie type G,(q) defined over finite fields GF(q) with q elements, as long as the Lie rank n of G,(q) is not too large. Since these computations require the symbolic manipulation of algebraic polynomials in one indeterminate q over a ring of algebraic integers, general computer algebra systems such as MAPLE (Char et al., 1985) and SCRATCHPAD II (Jenks, 1984) have recently found applications to the representation theory of finite groups of Lie type. Cohen's system LIE (Cohen, 1989) is also very useful.

In the following section, some of the noteworthy achievements are described. Furthermore, several other computational problems are mentioned, e.g. the computation of the value R~T,O(9) of a Deligne-Lusztig character at an element # of G,(q). Although there is a deterministic algorithm for the evaluation of R G T,0 at O, in general, the concrete computation presents tremendous difficulties.

Several sporadic simple groups have been constructed in terms of irreducible modular representations over small fields (Gorenstein, 1982). One of the most ingenious constructions is that used to construct the simple Janko group J4 by Benson et al. as a specific subgroup of the general linear group GL(112, 2) (Gorenstein, 1982). Variations of this method developed by Parker & Wilson (1990) and Gollan (1990) are described in section 3. The central problem here may be stated simply: How does one find explicit matrix generators for a given finite group G? Often they may be used to determine a set of defining relations satisfied by these generators.

The correspondences of Brauer, Green and Fong yield very powerful tools for the general modular representation theory of arbitrary finite groups G over fields F with characteristic $p > 0$ dividing the order IGI. However, many open problems remain in the subject. For example, no criteria are known for subgroups or indecomposable modules to be vertices or sources of simple FG-modules, respectively. In his book, Feit (1982, p. 121) writes: "Virtually nothing is known in this connection". It is therefore important to study numerous examples in great detail. For this, the new algorithms of Schneider (1990) for the computation of the vertices and sources of indecomposable modules M of dimension dim~M _< 300 are very helpful. They will be described in section 4. Effective use of these methods was made by Gollan (1990), who recently determined the sources of the simple 5-modular representations of the simple Tits-group 2F~(2)', His results are summarised in section 5, where, additionally, some related open problems on modular representations of groups with a T.I. Sylow p-subgroup are mentioned.

Most of the modular irreducible representations of the sporadic simple groups are now known. Parker's MEAT-AXE program and the MOC-computer package of Hiss et al. (in preparation) provide powerful tools for tackling this problem. Nevertheless, several hard cases still remain unsolved. In order to describe the irreducible modular and indecomposable projective representations of a p-block B with cyclic defect group, it is necessary to determine the Brauer tree of B. Although the book by Hiss & Lux (1989) contains most of the trees of the cyclic p-blocks of the simple sporadic groups, there are still some cases left open. In view of the strength of the theoretical results by Brauer and Dade, this is somewhat astonishing.

The source of the basic difficulties in all these computational problems, arising from modular representation theory, lies in the cost of performing computations with n x n matrices over finite fields GF(q). Here, the sizes of n and q impose serious restrictions for successful computer calculations. This is even more so for the symbolic computations which are necessary for the problems mentioned in section 2 concerning the determination of the character tables of some finite groups of Lie type.

It is therefore hoped that computer experiments will provide enough evidence for possible general mathematical results which then can be proved completely by means of theoretical methods.

Concerning terminology and notation we refer to the books by Benson (1984), Carter (1985), Feit (1982), Isaacs (1976) and the author's lecture notes (Michler, 1989).

2. CHARACTER TABLES OF FINITE GROUPS OF LIE TYPE

In 1907 Schur determined the complete character tables of the finite two-dimensional projective linear groups PGL(2, q) and the two-dlmensional special linear groups SL(2, q) defined over finite fields GF(q) with q elements. Almost 40 years later, Steinberg solved the same problem for the three- and four-dimensional linear groups. In 1973 Simpson and Frame treated the case of all three-dimensional unitary groups PSU(3, q2) and SU(3, q2). One year later, Chang and Ree determined the complete character table of the groups G2(q). Since then additional character tables have been calculated.

In 1976 Deligne and Lusztig published their famous paper "Representations of reductive groups over finite fields" in the Annals of Mathematics. In this paper they present general methods for the construction of families of irreducible characters of an arbitrary finite group of Lie type. Further advances have subsequently been made by Lusztig, and are presented in his book (Lusztig, 1984). In particular, the degrees of the irreducible representations of the finite groups of Lie type are now completely known. Furthermore, many character values can be computed. Since such a group has huge numbers of conjugacy classes in general, it is clear that its complete character table can only be determined if its Lie rank is small. Such a case has been dealt with by Deriziotis and the author (Deriziotis & Michler, 1987), where the character table of the Steinberg triality 3D4(q) is derived.

At present the character tables of the exceptional groups of Lie type F4(q), E,(q) for n = 6, 7 and 8 have not been completed for arbitrary q. In the Atlas (Conway et al., 1985) the character tables for F4(2) and 2E6(2) are given. Recently, Fischer (Bielefeld) has calculated the character table of E6(2). However, the tables for E7(2) and E8(2) are not yet known. For arbitrary q, the number of conjugacy classes for finite group of Lie type E,(q) with rank n is a polynomial in q with rational coefficients ate Z and degree n. The difficulty of determining the complete generic character table of ET(q) and Es(q) is indicated by:

Problem 1. Compute the rational coefficients ai of the polynomials

$$p_n(q) = q^n + \sum_{i=0}^{n-1} a_i q^i$$

describing the numbers of conjugacy classes of the simple groups E,(q) for n = 7, 8.

An understanding of the computational difficulties of this and some related partial problems requires the following notations and general results about finite groups of Lie type.

Every finite Chevalley or twisted group of Lie type considered here is the group G~ of fixed points of an endomorphism a of a simple connected algebraic group G defined over a finite field F= GF(q) with q=pm elements, where the prime number p>0 is the characteristic of F. The simply connected covering group of G is denoted by G~c and the adjoint group by G,d, the dual of Gsc. For further explanation, see Carter (1985).

Let T be a a-stable maximal torus of G. The maximal torus of Gr is a subgroup of the form Tr = G~ca T. The quotient We = (NG(T)/T)~ is the Weyl group of the maximal torus T~ of G,. It acts on Tr and its character group T~. The set Cw.(s) = {co e We I s ~ = s} i8 called the isotropy group of the semi-simple element s of T~. If Cw.(s) = 1, then s e T is said to be in general position. An element s of a maximal torus T is called regular if T coincides with the connected component C~(s) ~ of Co(s), where H ~ denotes the connected component of the group H. Elements s e G~ which are in general position, are regular; the converse is true in simply connected groups, but not in general.

Let W be the Weyl group of a fixed a-stable maximally split torus To of G. Then W = Nv(To)/To is uniquely determined by G up to G-conjugation. By Carter (1985, p. 84) the Gr classes of a-stable maximal tori T of G are in bijective correspondence with the tr-conjugacy classes of W. Two elements col, co2~ W are a-conjugate, if 0) 2 = coo)in(co) -1, for some co~ W.

A maximal torus T corresponding to the identity class of the Weyl group W is called a maximal split torus of G,. It is denoted by To.

Let G be a simply connected reductive algebraic group and a an endomorphism of G such that its group G, of fixed points is finite. If G has semi-simple Lie rank n, then G, has IZ~ ~ semi-simple conjugacy classes by Theorem 3.7.6. of Carter (1985). Here Z ~ denotes the connected centre of G. But in order to compute the total number of conjugacy classes of G,, it is necessary to find the precise answer to the following.

Problem 2. Let T be a maximal torus of Go. Determine the number of G~-conjugacy classes of regular elements having a representative in T.

Recently, Janiszczak computed these numbers for E6(q) and ET(q) using a machine. Similar computations for Ea(q) are in progress.

The Deligne-Lusztig generalised character Rr, o of a finite group G, of Lie type are defined in Chapter 7 of Carter (1985). Here, T denotes a a-stable maximal torus of the connected reductive group G, and 0 is a linear character of the finite abelian subgroup T~ of G,. The importance of the generalised characters Rr,o is best explained by Corollary 7.5.8 of Carter (1985): for every irreducible character Z of G~, there exists a generalised character Rr.o such that the inner product (Rr,0, Z)~ 0. Furthermore, Corollary 7.3.5. of Carter (1985) asserts that _. RT,O is an irreducible character of Go if 0 ~ Hom(T~, C) = 7"~ is in general position. Since such linear characters 0 ~ :Fo correspond to regular elements s of To under the duality between T~ and :F,, this consequence of the Deligne-Lusztig theory again demonstrates the importance of Problem 2 for the construction of irreducible characters of finite groups G, of Lie type.

It is well known that each element 9 e G, has a unique Jordan decomposition 9 = su = us, where u is a unipotent and s is a semi-simple element of Go. The Green function Qr is defined by Qr(u) = RT, l(u) for every unipotent element u ~ G~. According to Carter (1985), all its values QT(U) are rational integers.

If s is a semi-simple element of the reductive group G, then C~ denotes the connected centraliser of s in G. A very important character formula in the Deligne-Lusztig theory is the following:

Let 9~G~ have Jordan decomposition 9 = su = us, where s is semi-simple and u is unipotent. Then by (Carter 1985),

$$R_{T,\theta}(g) = \frac{1}{|C^0(s)_\sigma|} \sum_{\substack{x \in G_\sigma \\ x^{-1}sx \in T_\sigma}} \theta(x^{-1}sx) Q^{C^0(s)}_{xTx^{-1}}(u) \qquad (*)$$

Although formula (*) yields a complete algorithm for the evaluation of the DeligneLusztig character RT, o at an element g EG„ it is rather complicated to compute Rr, o(g), even in the case of small Lie rank n. It is first necessary to solve the following.

Problem 3. Find the values Qr(u) of the Green functions Qr at unipotent elements u E G.

In the books of Carter (1985) and Lusztig (1984), several methods for the solution of that problem have been given. Shoji (1982) has computed the Green polynomials for F~(q). Beynon & Spaltenstein (1984) solved the same problem for En(q), n = 6, 7 and 8. As can be seen from their paper, the computer has played an essential role. Using the computer algebra systems CAYLEY and REDUCE, Lambe & Srinivasan (1978) have computed the Green functions of some finite classical groups of type Cn (2 < n < 5), B, (3 < n < 5) and D~ (n = 4, 5) for q odd.

Having solved Problem 3 for a given group G, it is then still hard to find efficient algorithms for the solution of:

Problem 4. Find the values Rr, 0 (O), g r G„ using formula (*).

The case of semi-simple elements 9 = s does not involve the evaluation of the Green functions because of Proposition 7.5.3. of Carter (1985). Nevertheless, in order to be able to decide whether x-lsx ~ T„ it is necessary to have an algorithm for finding a complete system of double coset representatives x~ of T,. Also, the evaluation of the linear characters 0 requires extension of the present tools for computing with algebraic numbers. Thus far, these computational problems have only been solved for groups G,(q) with small Lie rank n~5.

In 1963 Ennola determined the complete character table of the unitary group U(3, q2). Ten years later, Simpson & Frame (1973) computed the character table of the special unitary group SU(3, q2). Using the computer algebra system MAPLE, Geck (1987) was able to find the restrictions of the irreducible characters of U(3, q2) to SU(3, q2) automatically. In fact, his method is a very efficient way to compute the character table of SU(3, q2). Since the character table of U(4, q2) is known, MAPLE can also be used to compute the character table of SU(4, q2).

3. CONSTRUCTIONS OF REPRESENTATIONS

In this section some methods for constructing group representations are reviewed. Throughout this section (F, R, S) will denote a splitting p-modular system for the finite group G (Feit, 1982 or Michler, 1989).

Definition 1. Let A ~ {F, R, S) and let U be a subgroup of G. If 1 v denotes the trivial AU-module, then M=(lv)G- -lv| is called the permutation module of G with respect to U.

Lemma 1.

(a) $\dim_A(1_U)^G = |G:U|$.
(b) *If $\kappa: G \to GL(\dim_A M, A)$ denotes the group representation of G afforded by $M = (1_U)^G$, then each $\kappa(g)$, $g \in G$, is a permutation matrix.*
(c) *The endomorphism rings of the permutation modules $M_F = 1_U \otimes_{FU} FG$, $M_R = 1_U \otimes_{RU} RG$ and $M_S = 1_U \otimes_{SU} SG$ satisfy the conditions:*

$$\mathrm{End}_{FG}(M_F) \cong F \otimes_R \mathrm{End}_{RG}(M_R),$$
$$\mathrm{End}_{SG}(M_S) \cong S \otimes_R \mathrm{End}_{RG}(M_R).$$

(d) $M_F \cong F \otimes_R M_R$ *and* $M_S \cong S \otimes_R M_R$.

In general, it may be difficult to find a permutation representation (iv) ~ of a group G by means of the Todd-Coxeter algorithm, see Cannon (1984). In order to determine the ring theoretical structure of the endomorphism ring EndF~[(lv) ~] of a permutation module (lv) G of a subgroup U · of G, it suffices to compute a complete system of representatives for the double cosets of U. For this easier task it is necessary to solve:

Problem 5. Find a deterministic algorithm for the computation of a complete system of representatives x for the U-U double cosets UxU of a subgroup U in the finite group G.

Apart from permutation modules, tensor product constructions are used. Theoretically, they are important because of the following classical result:

Theorem 1. (R. Brauer). Let A e {F, S} and let M be a faithful simple AG-module. Then every simple AG-module T is a composition factor of some tensor power $M^{\otimes n}$ *of M.*

In practice, tensor products are very difficult to split as they correspond to modules having fairly large dimension. Even wedge powers fM of FG-modules M are, in general, too large for concrete computer computations.

When Benson, Conway, Parker, Norton and Thackray constructed the simple Janko group J4 as a subgroup of GL(112, 2), they introduced a method for extending representations from pairs of proper subgroups of a group G to the whole group G. This technique has been slightly altered and generalised by Gollan (1989). In order to state his version of the result, the following notation is needed.

Deflnition 2. Let $\kappa : G \rightarrow GL(n, F)$ be an n-dimensional representation of the finite group G over the field F. Let U be a subgroup of G and $\kappa_{|U}$ the restriction of κ to U. Then x(U) is a subgroup of GL(n, F). The centraliser of tqu in GL(n, F) is defined by $C_{GL(n,F)}(\kappa_{|U}) = C_{GL(n,F)}(\kappa(U))$. In particular, $C_{GL(n,F)}(\kappa) = C_{GL(n,F)}(\kappa(G))$

Theorem 2. (Gollan), Let U1 = $\langle x_1, x_2, \ldots, x_r \rangle$ $U_2 = \langle y_1, y_2, \ldots, y_m \rangle$ be two subgroups of the finite group G such that G is generated by the x i and $y_j, 1 \le i \le r, 1 \le j \le m$. Let F be a splitting field for G wish characteristic p > 0 not dividing the order of $D = U_1 \cap U_2$. Suppose that for s = $1, 2, \kappa_s : U_s \rightarrow GL(n, F)$ is a representation of Us Xl and ~c z such that the restrictions $\kappa_{s|D}$ yield the same representation Of D, *i.e.* $\kappa_{1|D} = \kappa_{2|D}$. *Let*

$$k = |C_{GL(n,F)}(\kappa_{1|D}) : C_{GL(n,F)}(\kappa_1)|,$$

and let $T_1, T_2, \ldots, T_k \in GL(n, F)$ be a transversal for $C_{GL(n,F)}(\kappa_1)$ *in* $C_{GL(n,F)}(\kappa_{1|D})$. If κ_1 *and* κ_2 have a common extension $\kappa : G \rightarrow GL(n, F)$ to the whole group G, then there is a matrix $T \in \{T_u | 1 \le u \le k\}$ such that

$$\kappa(G) = \langle \kappa_1(x_1), \kappa_1(x_2), \ldots, \kappa_1(x_r), T\kappa_2(y_1)T^{-1}, \ldots, T\kappa_2(y_m)T^{-1} \rangle.$$

Furthermore, conjugation by T induces an isomorphism between $\kappa_{1|D}$ *and* $\kappa_{2|D}$.

Remark. The above theorem can also be used to show that two given subrepresentations $\kappa_s : U_s \to GL(n, \mathrm{F}),\ s = 1, 2,$ do not have a common extension x to the whole group G. In this case there does not exist a transformation matrix T such that

$$\kappa(G) = \langle \kappa_1(x_1), \kappa_1(x_2), \ldots, \kappa_1(x_r), T\kappa_2(y_1)T^{-1}, \ldots, T\kappa_2(y_m)T^{-1}\rangle.$$

Of course, the resulting computation for large groups may be very expensive.

Conversely, if it is known that κ_1 and κ_2 extend to an irreducible representation x of the whole group G, then it follows that there exist transformation matrices T such that the group x(G) is generated by the matrices $\kappa_1(x_i),\ 1 \leq i \leq r,$ and the matrices $T\kappa_2(y_j)T^{-1},\ 1 \leq j \leq m.$

Remark 2. Gollan (1990) has applied this theorem to construct the 26-dimensional 5-modular irreducible representation of the simple Tits group $G = {}^2\mathrm{F}_4(2)'$ having order $|G| = 2^{11}\ 3^3\ 5^2\ 13.$ Here, G contains two conjugacy classes of maximal subgroups U1 and U2 which are both isomorphic to the automorphism group of PSL(3, 3), i.e. Ul= PSL(3, 3) : 2, i = 1, 2. Their intersection $D = U_1 \cap U_2 \cong 3^{1+2} : D_8$ in the notation of the ATLAS, where 3 ~+2 denotes the extra-special 3-group of order 27 and exponent 3. In this example, the index $k = 24^4$.. For the details of the construction, see Gollan (1990),

Remark 3. In the original construction of J_4 by Norton and Parker, the hypothesis $p \nmid |U_1 \cap U_2|$ is not assumed. Nevertheless, they were able to describe the indecomposable summands of the restrictions of the representations κ_1 and κ_2 to $U_1 \cap U_2$. In general, the structure of these summands of the restrictions will not be known. For this reason Gollan considered semi-simple restrictions.

However, there are several other successful applications of the Cambridge method without the restrictive hypothesis. In particular, the Lyons group Ly has recently been constructed as a subgroup of GL(lll, 5) by Parker & Wilson (1990). Other applications are described in this paper.

4. INDUCTION AND RESTRICTION OF MODULAR REPRESENTATIONS

Throughout this section G will denote a finite group and F a finite field having characteristic p > 0 dividing the order]GI of G. By the Krull-Schmidt theorem every finitely generated FG-module M can be uniquely decomposed into indecomposable nonisomorphic FG-submodules M~ with multiplicities m~ such that

$$M \cong \sum_{i=1}^{s} \otimes m_i M_i,$$

where $m_i M_i$ denotes the direct sum of m_i copies of M_i.

Problem 6. Find an efficient algorithm for the decomposition of a finitely generated FG-module M into indecomposable direct summands M_i.

Of course, it suffices to find a complete set of orthogonal primitive idempotents e in the endomorphism ring $E = \mathrm{End}_{FG}(M)$ with sum equal to the identity element $1 \in E$. The basic step is:

Problem 7. Find a non-zero idempotent $e \neq 0$ in the endomorphism ring $E = \mathrm{End}_{FG}(M)$ of an FG-module M.

It turns out that M is indecomposable if and only if the identity endomorphism of M is the only non-zero idempotent in $E = \mathrm{End}_{FG}(M)$.

Schneider (1990) has recently given probabilistic algorithms for Problems 6 and 7. They work very well as long as $\dim_F M \leq 300$. They have been installed in the computer algebra system CAYLEY. The socle series of an FG-module M is a series

$$0 = \mathrm{soc}_0(M) < \mathrm{soc}_1(M) < \mathrm{soc}_2(M) < \ldots < \mathrm{soc}_s(M) = M$$

of submodules of M such that $\mathrm{soc}_i(M)/\mathrm{soc}_{i-1}(M)$ is the largest semi-simple submodule of $M/\mathrm{soc}_{i-1}(M)$ for $1 \leq i \leq s$. The integer s is called the socle length of M.

In practice, Parker's Meataxe may sometimes be used to compute the soele series of an FG-module M. Since the submodules $\mathrm{soc}_i(M)$ are invariant under all endomorphisms of M, the socle series of M may provide an efficient way to solve Problems 6 or 7 in specific cases.

In modular representation theory, Green's theory of vertices and sources plays an important role. A vertex of an Jndecomposable FG-module M is a p-subgroup S of smallest order such that the FG-module $M\downarrow_S\uparrow^G$ has a direct summand isomorphic to M. In fact, $S = vx(M)$ is uniquely determined by M up to G-conjugacy. An indecomposable FS-module Q is a source of M if M is isomorphic to a direct summand of the induced FG-module $Q\uparrow^G = Q \otimes_{FS} FG$.

Problem 8. Let M be an indecomposable FG-module. Find deterministic algorithms for the determination of the vertices vx(M) and sources Q of M.

Schneider (1990) has developed algorithms for Problem 8 which are effective for modules M such that $\dim_F M \leq 300$. The algorithms developed by Schneider include an algorithm for the computation of indempotents in the endomorphism ring $E = \mathrm{End}_{FG}(M)$ and an algorithm WRDSOL that returns any element of a group G as word in the given set $\{g_i | 1 \leq i \leq k\}$ of generators of G. This latter algorithm enabled him to implement very efficient algorithms in CAYLEY for the induction and restriction of representations from and to subgroups U of G, respectively. They are very useful for computations with matrix generators of a group.

For an idempotent $e = e^2 \neq 0$ of the group algebra FG of a finite group G, the projective FG-module eFG has endomorphism ring $\mathrm{End}_{FG}(eFG) \cong eFGe$. Furthermore, for each right FG-module M, there is a right eFGe-module Me. Of course, it is only necessary to choose idempotents e such that the condensed eFGe-module Me has dimension $\dim_F Me < \dim_F M$. If by means of the MEAT-AXE or other methods it is possible to find a composition series or the socle of the eFGe-module Me, then it is sometimes possible to obtain information about the composition series, socle or composition factors of the FG-module M. Recently, Ryba (1990) has successfully applied this method of condensation by constructing a new irreducible 2-modular G2(3)-module of dimension 378. He uses an idempotent e of the form

$$e = \frac{1}{|H|} \sum_{h \in H} h, \quad \text{where } (|H|, \mathrm{char}(F) = p) = 1.$$

Furthermore, Ryba assumes that the p'-group H acts monomially on the FG-module M, and that $\dim_F M < 1000|H|$. In his case, the module M is an nth wedge power, $\wedge^n W$, of a fairly small FG-module W. According to Ryba, the new condensation program can be applied to a large class of representations M with $\dim_F M \leq 10^6$. Nevertheless, this method is also demanding. Ryba (1990) points out that the user of the condensation program must solve the following preliminary.

Problem 9. Let $g_1, g_2, \ldots, g_r$ be a set of generators of the finite group G represented on the FG-module M by the matrices $\kappa(g_i) \in GL(\dim_F M, F)$, $1 \leq i \leq r$. Suppose that the generators $h_1, h_2, \ldots, h_t$ of the p'-subgroup H can be written as words in the generators gi. Then it is necessary to find group elements $v_1, v_2, \ldots, v_s \in G$ written as words in the gt such that $\{y_1 = ev_1e,\ y_2 = ev_2e, \ldots, y_s = ev_se\}$ generate the F-algebra eFGe, where

$$e = \frac{1}{|H|} \sum_{h \in H} h.$$

The (eFGe)-module structure of Me is then determined by the matrices $\kappa(y_1, \kappa(y_2), \ldots, \kappa(y_s)$.

In the particular case, $G = G_2(3)$, Ryba was able to overcome the difficulty of Problem 9 using a clever trick.

A classical problem of Clifford theory is concerned with the following question: Suppose that U is a normal subgroup of G, and that M is a simple FU-module which is inertial in G. How can the irreducible components of $M^G = M \otimes_{FU} FG$ be efficiently determined? As is well known, this problem is equivalent to that of finding all the irreducible projective representations of G/U (Isaacs, 1976). Recently, Conlon (1990) has given efficient algorithms for the computation of the degrees of the irreducible projective representations of a finite solvable group. These methods may also be applied to the above induction problem provided G/U is a solvable group. Since the construction of the projective representations of a group G depends on the Schur multiplier of G, Holt's algorithm (1984) for calculating the p-primary part of the Schur multiplier of a finite permutation group is relevant here.

5. GROUPS WITH A T.I. SYLOW P-SUBGROUP

Blau and the author have studied the modular representations of a finite group G with a T.I. Sylow p-subgroup S, i.e. $S \cap xSx^{-1} = 1$ for all $x \notin N_G(S)$ (Blau & Michler, 1990). For these groups G, the Brauer's height zero conjecture and the Alperin-McKay conjecture in block theory were verified. Unfortunately, the results depend on the classification of the finite groups (Gorenstein, 1982). Furthermore, it was necessary to calculate the character tables of the Sylow normalisers occurring in the possible minimal counter-examples. Most of these tables were calculated by means of the character table algorithms contained in CAYLEY. However, even for these groups G with a T.I. Sylow p-subgroup, the authors were not able to answer Brauer's famous problem which asserts that the number k(B) of irreducible characters of a p-block B of a finite group G with defect group 6(B) =oD is bounded by k(B) < IDI. In fact, the open case where S = D, a normal Sylow p-subgroup of G, is extremely hard. The results of Blau & Michler (1990) suggest the following:

The results of Blau & Michler (1990) suggest the following:

Problem 10. Let G be a finite group with a T.I. Sylow p-subgroup S. Let M be a simple FG-module with vertex $vx(M) =_G S$ and Green correspondent $f(M)$ in $H = N_G(S)$ defined by

$$M{\downarrow_H} = f(M) \oplus \text{projective},$$

where f(M) is the unique non-projective FH-module of the restriction $M{\downarrow_H}$. Is the socle $\mathrm{soc}_1[f(M)]$ of $f(M)$ multiplicity free?

In order to show the strength of the algorithms described in the previous sections, the following results of Gollan which also support Problem 10 are now explained (Gollan, 1990).

Let $G = {}^2F_4(2)'$ be the simple Tits group. Its Sylow 5-subgroup S is a T.I. set. By Hiss (1986) the principal 5-block B of G has simple FG-modules of dimensions 1, 26, 27, 78, 109, 351, 460 and 593 over the field F = GF(25) with 25 elements, where 1 denotes the trivial FG-module. Here, the modules of dimensions 26, 27 and 351 are not isomorphic to their respective duals 26*, 27* and 351". Furthermore, there are non-isomorphic algebraic conjugate modules 109', 460' and 593'. As mentioned in the ATLAS (Conway et al., 1985), the Sylow 5-normaliser $H \cong (\mathbb{Z}_5 \times \mathbb{Z}_5):4A_4$. Hence, H has the following simple FH-modules named according to their dimensions:

$$I, 1, 1_a, 1_a^*, 1_b, 1_b^*, 2_a, 2_a^*, 2_b, 2_b^*, 2_c, 2_c^*, 3_a, 3_b.$$

By Gollan (1989), the Green correspondents f(n) of the simple FG-modules n have the following socle series:

$f(26) = 1_a^*,\qquad f(26^*) = 1_a$

$$f(27) = \begin{array}{ccccc} & & 1 & & \\ & & 2_a & & \\ & 3_b & & 3_b & \\ 2_a^* & & 2_b & & 2_c^* \\ 1_a & & 1_a^* & & 3_a \\ & 2_b^* & & 2_c & \\ & & 3_b & & \end{array},\qquad f(27^*) = \begin{array}{ccccc} & & 3_b & & \\ & 2_b & & 2_c^* & \\ 1_a & & 1_a^* & & 3_a \\ & 2_b^* & & 2_c & \\ & & 3_b & & \\ & 2_a & & 2_a^* & \\ & 1 & & 3_b & \end{array}$$

$$f(78) = \begin{array}{ccccc} & & 2_a^* & & \\ & & 3_a & & \\ 2_a & & 2_b^* & & 2_c \\ 1_b & & 1_b^* & & 3_b \\ & 2_b & & 2_c^* & \\ & \mathrm{I} & & 3_a & \\ & 2_a & & 2_a^* & \end{array}$$

$$f(109) = \begin{array}{c} 1_b \\ 2_b \\ 3_a \\ 2_b^* \\ 1_b^* \end{array},\qquad f(109') = \begin{array}{c} 1_b^* \\ 2_c^* \\ 3_a \\ 2_c \\ 1_b \end{array}$$

$$f(351) = \begin{array}{ccccccccccccccc} & & & & & & & 3_a & & & & & & & \\ & & & & & 2_a & & 2_b^* & & 2_c & & & & & \\ & & & \mathrm{I} & & 1_b & & 1_b^* & & 3_b & & 3_b & & & \\ 2_a^* & & 2_a^* & & 2_b & & 2_b & & 2_b & & 2_c^* & & 2_c^* & & 2_c^* \\ & 1 & & 1_a & & 1_a^* & & 3_a & & 3_a & & 3_a & & 3_a & \\ & & & \mathrm{I} & & 1_b & & 1_b^* & & 3_b & & 3_b & & & \\ & & & & & 2_a^* & & 2_b & & 2_c^* & & & & & \end{array}$$

$$f(351^*) = \begin{array}{ccccccccccccc} & & & & 2_a & & 2_b^* & & 2_c & & & & \\ & & \mathrm{I} & & 1_b & & 1_b^* & & 3_b & & 3_b & & \\ & 2_a^* & & 2_a^* & & 2_b & & 2_b & & 2_c^* & & 2_c^* & \\ 1 & & 1_a & & 1_a^* & & 3_a & & 3_a & & 3_a & & 3_a \\ & 2_a & & 2_a & & 2_b^* & & 2_b^* & & 2_c & & 2_c & \\ & & \mathrm{I} & & 1_b & & 1_b^* & & 3_b & & 3_b & & \\ & & & & 2_a^* & & 2_b & & 2_c^* & & & & \\ & & & & 2_b^* & & 2_c & & 3_a & & & & \end{array}$$

$$f(460) = \begin{array}{cccccc} & & 2_a^* & 2_b & & \\ & 1 & 1_a & 3_a & 3_a & \\ 2_a & 2_a & 2_b^* & 2_b^* & 2_c & 2_c \\ \mathrm{I} & 1_b & 1_b^* & 3_b & 3_b & 3_b \\ 2_a^* & 2_a^* & 2_b & 2_b & 2_c^* & 2_c^* \\ & 1 & 1_a^* & 3_a & 3_a & \\ & & 2_a & 2_b^* & & \end{array}$$

$$
f(460') = \begin{array}{cccccc}
 & & 2^*_a & 2^*_c & & \\
 & 1 & 1^*_a & 3_a & 3_a & \\
2_a & 2_a & 2^*_b & 2^*_b & 2_c & 2_c \\
I & 1_b & 1^*_b & 3_b & 3_b & 3_b \\
2^*_a & 2^*_a & 2_b & 2_b & 2^*_c & 2^*_c \\
 & 1 & 1_a & 3_a & 3_a & \\
 & & & 2_a & 2_c &
\end{array}
$$

$$
f(593) = \begin{array}{ccc}
 & 3_a & \\
2_a & & 2^*_b \\
I & & 3_b \\
2^*_a & & 2_b \\
 & 3_a &
\end{array}, \qquad
f(593') = \begin{array}{ccc}
 & 3_a & \\
2_a & & 2_c \\
I & & 3_b \\
2^*_a & & 2^*_c \\
 & 3_a &
\end{array}
$$

Remark 4. Except for the cases (460, 460'), Gollan (1990) has shown that the restriction of the Green correspondent to the Sylow p-subgroup S of G is indecomposable. Thus, the restrictions of the Green correspondentsf(n) to S are the sources of the simple FG-modules n. In the cases (460, 460'), this restriction off(n) decomposes into four summands.

Remark 5. The Sylow 5-subgroup S of the McLaughlin group McL is a T.I. set. With the present methods it is possible to answer Problem 10 for McL. This will be completed in Essen in due course.

If the assertion of Problem 10 is also verified for the McLaughlin group McL, then by Blau & Michler (1990) the only possible counter-examples G to Problem 10 have to involve the simple Janko group J4 as a composition factor. Since the dimensions of the non-trivial simple FJ_4-modules over the field $F = GF(11)$, all exceed 1332, the only manageable candidate is the simple module of dimension 1333. Of course, the method of section 3 will lead to a new construction of ,/4 over GF(ll). The restriction of this representation to the Sylow 11-normaliser of J4 can then be computed by the methods of section 4. At present it cannot be decided whether this example will provide a counterexample or even more evidence for an affirmative answer to the question posed as Problem 10.

The author thanks J. Cannon and the referees for many helpful suggestions improving the presentation of this article.

REFERENCES

1. Benson, D. (1984). Modular representation theory: New trends and methods. In Lecture Notes in Math. 1081. Heidelberg: Springer-Verlag.

2. Beynon, W. M., Spaltenstein, N. (1984). Green functions of finite Chevalley groups of E, (n = 6, 7, 8). J. Algebra 88, 584-614.

3. Blau, H. L, Miehler, G. O. (1990). Modular representation theory of finite groups with T.I. Sylow p-subgroups. Transact. Amer. Math. Soc. (to be published).

4. Cannon, J. J. (1984). An introduction to the group theory language CAYLEY. In Computational Group Theory (Atkinson, M. D., ed.), pp. 145-183. London: Academic Press.

5. Carter, R, W. (1972). Simile Groups of Lie Type. New York: Wiley.

6. Carter, R. W. (1985). Finite Groups of Lie Type: Conjugacy Classes and Complex Characters. New York: Wiley.

7. Char, J. B., Geddes, K. O., Gonnet, G. H., Watt, S. M. (1985). First Leaves: A Tutorial Introduction to MAPLE and MAPLE REFERENCE MANUAL, 4th edn. Canada: University of Waterloo.

8. Cohen, A. M. (1989). Lie Manual. Center of Mathematics and Computer Science, Amsterdam.

9. Conlon, S. B. (I990). Computing modular and projective character degrees of soluble groups. J. Symbolic Comp. 9, 551-570.

10. Conway, 3'. H., Curtis, R. T., Norton, S. P., Parker, R. A., Wilson, R. A. (1985). Atlas of Finite Groups. Oxford: Clarendon Press.

11. Deriziotis, D. I. & Michler, G. O. (1987). Character table and blocks of finite simple triality groups 3D4(q). Transact. Amer. Math. Soc. 303, 39-70.

12. Feit, W. (1982). The Representation Theory of Finite Groups. Amsterdam: North-Holland.

13. Geck, M. (1987). Eine Anwendung yon MAPLE in der Darstellungstheorie der unitfiren Gruppen. Diplomarbeit, RWTH Aachen.

14. Gollan, H. W. (1990). The 5-modular Irreducible Representations of 2F4(2)', Their Green Correspondents and Sources. Math. Comp. (to appear).

15. Gorenstein, D. (1982). Finite Simple Groups, an Introduction to The& Classification. New York: Plenum Press.

16. HiB, G. (1986). The modular characters of the Tits simple groups and its automorphism group. Comm. Alg. 14, 125-154.

17. HiB, G., Lux, K., Parker, R. A. (1986). MOC-Programmsystem far die Berechnung yon Brauer-Charakteren. RWTH Aachen.

18. HiB, G., Lux, K. (1989). Brauer Trees of Sporadic Groups. Oxford University Press.

19. Holt, D. F. (1984). The calculation of the Schur multiplier of a permutation group. In Computational Group Theory (Atkinson, M. D., ed.), pp. 307-319. London: Academic Press.

20. Humphreys, J. E. (1972). Introduction to Lie Algebras and Representation Theory. Heidelberg: Springer.

21. Isaacs, J. M. (1976). Character Theory of Finite Groups. New York: Academic Press.

22. Jenks, R. (1984). Scratehpad II. Mathematics Department, IBM Research Center, Yorktown Heights.

23. Lambe, L., Srinivasan, B. (1987). A Computation of Green Functions for Some Classical Groups. Preprint.

24. Lusztig, G. (1984). Characters of Reductive Groups over a Finite Field. Princeton: Princeton University Press.

25. Michler, G. O. (1989). Ring theoretical and computational methods in group representation theory. Vorlesungen Fachbereich Mathematik der Universit/it Essen, Heft 18, Essen.

26. Neubfiser, 3.., Pahlings, H., Plesken, W. (1984). CAS; Design and use of a system for handling of characters of finite groups. In Computational Group Theory (Atkinson, M. D., ed.), pp. 195-237. London: Academic Press.

27. Parker, R. A., Wilson, R. A. (1990). The computer construction of matrix representations of finite groups over finite fields. J. Symbolic Comp. 9, 583-590.

28. Ryba, A. J'. E. (1990). Computer condensation of modular representations. 3". Symbolic Comp. 9, 591-600.

29. Schneider, G. 3.. A. (1990). Computing with endomorphism rings of modular representations. J. Symbolic Comp. 9, 607-636.

30. Shoji, T. (1982). On the Green polynomials of a Chevally group of type F 4. Comm. Algebra 10, 505-543.

31. Simpson, W. A., Frame, 3'. S. (1973). The character tables for SL(3,q), SU(3,q2), PSL(3,q), PSU(3,q2). Canad. J. Math. 25, 82-107.

2

TOPOLOGICAL METHODS IN MODULI THEORY

ABSTRACT

One of the main themes of this long article is the study of projective varieties which are K(H,1)'s, i.e. classifying spaces BH for some discrete group H. After recalling the basic properties of such classifying spaces, an important class of such varieties is introduced, the one of Bagnera–de Franchis varieties, the quotients of an Abelian variety by the free action of a cyclic group. Moduli spaces of Abelian varieties and of algebraic curves enter into the picture as examples of rational K(H,1)'s, through Teichmüller theory. The main trhust of the paper is to show how in the case of K(H,1)'s the study of moduli spaces and deformation classes can be achieved through by now classical results concerning regularity of classifying maps. The Inoue type varieties of Bauer and Catanese are introduced and studied as a key example, and new results are shown. Motivated from this study, the moduli spaces of algebraic varieties, and especially of algebraic curves with a group of automorphisms of a given topological type are studied in detail, following new results by the author, Michael Lönne and Fabio Perroni. Finally, the action of the absolute Galois group on the moduli spaces of such K(H,1) varieties is studied. In the case of surfaces isogenous to a product, it is shown how this yields a faifhtul action on the set of connected components of the moduli space: for each Galois automorphism of order different from 2 there is an algebraic surface S such that the Galois conjugate surface of S has fundamental group not isomorphic to the one of S.

KEYWORDS

Moduli spaces, Projective varieties, Classifying spaces, Group cohomology, Group homology, Symmetry marked moduli spaces, Group of automorphisms, Bagnera–de Franchis varieties, Absolute Galois group

1 INTRODUCTION

The interaction of algebraic geometry and topology has been such, in the last three centuries, that it is often difficult to say when does a result belong to one discipline or to the other, the archetypical example being the Bézout theorem, first conceived through a process of geometrical degeneration (algebraic hypersurfaces degenerating to union of hyperplanes), and later clarified through topology and through algebra.

This 'caveat' is meant to warn the reader that a more appropriate title for the present survey article could be: 'Some topological methods in moduli theory, and from the personal viewpoint, taste and understanding of the author'. In fact, many topics are treated, some classical and some very recent, but with a choice converging towards some well defined research interests.

I considered for some time the tempting and appealing title 'How can the angel of topology live happily with the devil of abstract algebra', paraphrasing the motto by Hermann Weyl.1

The latter title would have matched with my personal philosophical point of view: while it is reasonable that researchers in mathematics develop with enthusiasm and dedication new promising mathematical tools and theories, it is important then that the accumulated knowledge and cultural wealth (the instance of topology in the twentieth century being a major one) be not lost afterwards. This wealth must indeed not only be invested and exploited, but also further developed by addressing problems in other fields, problems which often raise new and fascinating questions. In more down to earth words, the main body of the article is meant to be an invitation for algebraic geometers to use more classical topology. This invitation is not new, see for istance the work of Atiyah and Bott on the moduli spaces of vector bundles on curves [15]; but explains the structure of the article which is, in a sense, that of a protracted colloquium talk, and where we hope that also topologists, for which many of these notions are well known, will get new kicks coming from algebraic geometry, and especially moduli theory.

In this article we mostly consider moduli theory as the fine part of classification theory of complex varieties: and we want to show how in some lucky cases topology helps also for the fine classification, allowing the study of the structure of moduli spaces: as we have done quite concretely in several papers [30–33, 35–37].

We have already warned the reader about the inhomogeneity of the level assumed in the text: usually many sections start with very elementary arguments but, at a certain point, when we deal with current problems, the required knowledge may raise considerably.

Let us try to summarize the logical thread of the article.

Algebraic topology flourished from some of its applications (such as Brouwer's fixed point theorem, or the theorem of Borsuk–Ulam) inferring the non existence of certain continuous maps from the observation that their existence would imply the existence of homomorphisms satisfying algebraic properties which are manifestly impossible to be verified.

Conversely, the theory of fibre bundles and homotopy theory give a topological incarnation of a group Gthrough its classifying space BG. The theory of classifying spaces translates then group homomorphisms into continuous maps to classifying spaces. For instance, in algebraic geometry, the theory of Albanese varieties can be understood as dealing with the case where G is free abelian and the classifying maps are holomorphic.

For more general G, an important question is the one of the regularity of these classifying maps, such as harmonicity, addressed by Eells and Sampson, and their complex analyticity addressed by Siu and others. These questions, which were at the forefront of mathematical research in the last 40 years, have powerful applications to moduli theory.

After a general introduction directed towards a broader public, starting with classical theorems by Zeuthen-Segre and Lefschetz, proceeding to classifying spaces and their properties, I shall concentrate on some classes of projective varieties which are classifying spaces for some group, providing several explicit examples. I discuss then locally symmetric varieties, and at a certain length the quotients of Abelian varieties by a cyclic group acting freely, which are here called Bagnera–De Franchis varieties.

At this point the article becomes instructional, and oriented towards graduate students, and several important topics, like orbifold fundamental groups, Teichmüller spaces, moduli spaces of curves, group cohomology and homology are treated in detail (and a new proof of a classical theorem of Hopf is sketched).

Then some applications are given to concrete problems in moduli theory, in particular a new construction of surfaces with pg=q=1 is given.

The next section is devoted to a preparation for the rigidity and quasi-rigidity properties of projective varieties which are classifying spaces (meaning that their moduli spaces are completely determined by their topology); in the section are recalled the by now classical results of Eells and Sampson, and Siu's results about complex analyticity of harmonic maps, with particular emphasis on bounded domains and locally symmetric varieties.

Other more elementary results, based on Hodge theory, the theorem of Castelnuovo-De Franchis, and on the explicit constructions of classifying spaces are explained in detail because of their importance for Kähler manifolds. We then briefly discuss Kodaira's problem and Voisin's counterexamples, then we dwell on fundamental groups of projective varieties, and on the Shafarevich conjecture.

Afterwards we deal with several concrete investigations of moduli spaces, which in fact lead to some group theoretical questions, and to the investigation of moduli spaces of varieties with symmetries.

Some key examples are: varieties isogenous to a product, and the Inoue-type varieties introduced in recent work with Ingrid Bauer: for these the moduli space is determined by the topological type. I shall present new results and open questions concerning this class of varieties.

In the final part, after recalling basic results on complex moduli theory, we shall also illustrate the concept of symmetry marked varieties and their moduli, discussing the several reasons why it is interesting to consider moduli spaces of triples (X,G,α) where X is a projective variety, G is a finite group, and α is an effective action of G on X. If X is the canonical model of a variety of general type, then G is acting linearly on some pluricanonical model, and we have a moduli space which is a finite covering of a closed subspace MG of the moduli space.

In the case of curves we show how this investigation is related to the description of the singular locus of the moduli space Mg (for instance of its irreducible components, see [129]), and of its compactification ¯Mg (see [113]).

In the case of surfaces there is another occurrence of Murphy's law, as shown in my joint work with Ingrid Bauer [33]: the deformation equivalence for minimal models S and for canonical models differs drastically (nodal Burniat surfaces being the easiest example). This shows how appropriate it is to work with Gieseker's moduli space of canonical models of surfaces.

In the case of curves, there are interesting relations with topology. Moduli spaces M$g(G)$ of curves with a group G of automorphisms of a fixed topological type have a description by Teichmüller theory, which naturally leads to conjecture genus stabilization for rational homology groups. I will then describe two equivalent descriptions of the irreducible components of M$g(G)$, surveying known irreducibility results for some special groups. A new fine homological invariant was introduced in our joint work with Lönne and Perroni: it allows to prove genus stabilization in the ramified case, extending a beautiful theorem due to Livingston [272] and Dunfield and Thurston [141], who dealt with the easier unramified case.

Another important application is the following one, in the direction of arithmetic: in the 60's Serre [336] showed that there exists a field automorphism σ in the absolute Galois group Gal(¯Q/Q), and a variety Xdefined over a number field, such that X and the Galois conjugate variety $X\sigma$ have non isomorphic fundamental groups, in particular they are not homeomorphic.

In a joint paper with I. Bauer and F. Grunewald we proved a strong sharpening of this phenomenon discovered by Serre, namely, that if σ is not in the conjugacy class of the complex conjugation then there exists a surface (isogenous to a product) X such that X and the Galois conjugate variety $X\sigma$ have non isomorphic fundamental groups.

In the end we finish with an extremely quick mention of several interesting topics which we do not have the time to describe properly, among these, the stabilization results for the cohomology of moduli spaces and of arithmetic varieties.

2 PREHISTORY AND BEYOND

The following discovery belongs to the 19-th century: consider the complex projective plane P2 and two general homogeneous polynomials

$F,G \in \mathbb{C}[x_0,x_1,x_2]$ of the same degree d. Then F, G determine a **linear pencil** of curves C_λ, $\forall \lambda=(\lambda_0,\lambda_1) \in \mathbb{P}^1$,

$$C_\lambda := \left\{x = (x_0, x_1, x_2) \in \mathbb{P}^2 | \lambda_0 F(x) + \lambda_1 G(x) = 0\right\}.$$

One sees that the curve C_λ is singular for exactly $\mu=3(d-1)^2$ values of λ, as can be verified by an elementary argument which we now sketch.

In fact, x is a singular point of some C_λ iff the following system of three homogeneous linear equations in $\lambda=(\lambda_0,\lambda_1)$ has a nontrivial solution:

$$\lambda_0 \frac{\partial F}{\partial x_i}(x) + \lambda_1 \frac{\partial G}{\partial x_i}(x) = 0, \quad \forall i = 0, 1, 2.$$

By generality of F, G we may assume that the curves $C_0:=\{x|F(x)=0\}$ and $C_1:=\{x|G(x)=0\}$ are smooth and intersect transversally (i.e., with distinct tangents) in d^2 distinct points; hence if a curve of the pencil C_λ has a singular point x, then we may assume that for this point we have $F(x)\neq 0 \neq G(x)$, and then λ is uniquely determined.

If now $\frac{\partial F}{\partial x_0}$ and $\frac{\partial G}{\partial x_0}$ do not vanish simultaneously in x, then the above system has a nontrivial solution if and only if

$$\left[\frac{\partial G}{\partial x_0} \cdot \frac{\partial F}{\partial x_i} - \frac{\partial F}{\partial x_0} \cdot \frac{\partial G}{\partial x_i}\right](x) = 0, \quad i = 1, 2.$$

By the theorem of Bézout (see [369]) the above two equations have $(2(d-1))^2=4(d-1)^2$ solutions, including among these the $(d-1)^2$ solutions of the system of two equations $\frac{\partial F}{\partial x_0}(x) = \frac{\partial G}{\partial x_0}(x) = 0$.

One sees that, for F, G general, there are no common solutions of the system

$$\frac{\partial F}{\partial x_0}(x) = \frac{\partial G}{\partial x_0}(x) = \left[\frac{\partial G}{\partial x_1} \cdot \frac{\partial F}{\partial x_2} - \frac{\partial F}{\partial x_1} \cdot \frac{\partial G}{\partial x_2}\right](x) = 0,$$

hence the solutions of the above system are indeed

$$3(d-1)^2 = 4(d-1)^2 - (d-1)^2.$$

It was found indeed that, rewriting $\mu=3(d-1)^2=d^2+2d(d-3)+3$, the above formula generalizes to a beautiful formula, valid for any smooth algebraic surface *S*, and which is the content of the so-called theorem of Zeuthen-Segre; this goes as follows: observe in fact that *d*2 is the number of points where the curves of the pencil meet, while the genus *g* of a plane curve of degree *d* equals $\frac{(d-1)(d-2)}{2}$.

Theorem 1 (Zeuthen-Segre, classical) Let *S* be a smooth projective surface, and let *C*λ, λ∈P1 be a linear pencil of curves of genus *g* which meet transversally in δ distinct points. If μ is the number of singular curves in the pencil (counted with multiplicity), then

$$\mu - \delta - 2(2g - 2) = I + 4,$$

where the integer *I* is an invariant of the algebraic surface, called Zeuthen-Segre invariant.

Here, the integer δ equals the self-intersection number *C*2 of the curve *C*, while in modern terms the number 2*g*−2=*C*2+*KS*·*C*, *KS* being the divisor (zeros minus poles) of a rational differential 2-form.

In particular, our previous calculation shows that for P2 the invariant *I*=−1.

The interesting part of the discovery is that the integer *I*+4 is not only an algebraic invariant, but is indeed a topological invariant.

Indeed, for a compact topological space *X* which can be written as the disjoint union of locally closed sets *Xi*, *i*=1,...*r*, homeomorphic to an Euclidean space R*ni*, one can define

$$e(X) := \sum_{i}^{r} (-1)^{n_i},$$

and indeed this definition is compatible with the more abstract definition

$$e(X) = \sum_{j=0}^{\dim(X)} (-1)^j \operatorname{rank} H_j(X, \mathbb{Z}).$$

For example, the plane $\mathbb{P}^2 = \mathbb{P}^2_{\mathbb{C}}$ is obtained from a point attaching $\mathbb{C}^2 = \mathbb{R}^4$, hence $e(\mathbb{P}^2) = 3$, hence *e*(P2)=3and we verify that *e*(P2)=*I*+4.

While for an algebraic curve *C* of genus *g* its topological Euler–Poincaré characteristic, for short Euler number, equals *e*(*C*)=2−2*g*, since *C* is obtained as the disjoint union of one point, $2g$ arcs, and a 2-disk (think of the topological realization as the quotient of a polygon with $4g$ sides).

The Euler Poincaré characteristic is multiplicative for products:

$$e(X \times Y) = e(X) \cdot e(Y),$$

and more generally for fibre bundles (a concept we shall introduce in the next section), and accordingly there is a generalization of the theorem of Zeuthen-Segre:

Theorem 2 (Zeuthen-Segre, modern) Let S be a smooth compact complex surface, and let $f:S \to B$ be a fibration onto a projective curve B of genus b (i.e., the fibres $f^{-1}(P)$, $P \in B$, are connected), and denote by g the genus of the smooth fibres of f. Then

$$e(S) = (2b - 2)(2g - 2) + \mu,$$

where $\mu \geq 0$, and $\mu = 0$ if and only if all the fibres of f are either smooth or, in the case where $g = 1$, a multiple of a smooth curve of genus 1.

The technique of studying linear pencils turned out to be an invaluable tool for the study of the topology of projective varieties. In fact, Solomon Lefschetz in the beginning of the 20-th century was able to describe the relation holding between a smooth projective variety $X \subset PN$ of dimension n and its hyperplane section $W = X \cap H$, where H is a general linear subspace of codimension 1, a hyperplane.

The work of Lefschetz deeply impressed the Italian algebraic geometer Guido Castelnuovo, who came to the conclusion that algebraic geometry could no longer be carried over without the new emerging techniques, and convinced Oscar Zariski to go on setting the building of algebraic geometry on a more solid basis. The report of Zariski [379] had a big influence and the results of Lefschetz were reproven and vastly extended by several authors: they say essentially that homology and homotopy groups of real dimension smaller than the complex dimension n of X are the same for X ands its hyperplane section W.

In my opinion the nicest proofs of the theorems of Lefschetz are those given much later by Andreotti and Frankel [4, 5].

Theorem 3 Let X be a smooth projective variety of complex dimension n, let $W=X\cap H$ be a smooth hyperplane section of X, and let further $Y=W\cap H'$ be a smooth hyperplane section of W.

First Lefschetz' theorem: the natural homomorphism $Hi(W,Z)\to Hi(X,Z)$ is bijective for $i<n-1$, and surjective for $i=n-1$; the same is true for the natural homomorphisms of homotopy groups $\pi i(W)\to\pi i(X)$(the results hold more generally, see [288], p. 41, even if X is singular and W contains the singular locus of$X$).

Second Lefschetz' theorem: The kernel of $Hn-1(W,Z)\to Hn-1(X,Z)$ is the subgroup Van$Hn-1(W,Z)$generated by the vanishing cycles, i.e., those cycles which are mapped to 0 when W tends to a singular hyperplane section $W\lambda$ in a pencil of hyperplane sections of X.

Generalized Zeuthen-Segre theorem: if μ is the number of singular hyperplane sections in a general linear pencil of hyperplane sections of X, then

$$e(X) = 2e(W) - e(Y) + (-1)^n \mu.$$

Third Lefschetz' theorem or Hard Lefschetz' theorem:

The first theorem and the universal coefficients theorem imply for the cohomology groups that $Hi(X,Z)\to Hi(W,Z)$ is bijective for $i<n-1$, while $Hn-1(X,Z)\to Hn-1(W,Z)$ is injective. Defining Inv$Hn-1(W,Z)$ as the Poincaré dual of the image of $Hn-1(X,Z)$, then we have a direct sum decomposition (orthogonal for the cup product) after tensoring with Q:

$$H_{n-1}(W,\mathbb{Q}) = Inv\, H_{n-1}(W,\mathbb{Q}) \oplus Van\, H_{n-1}(W,\mathbb{Q}).$$

Equivalently, the operator L given by cup product with the cohomology class $h\in H2(X,Z)$ of a hyperplane, $L{:}Hi(X,Z)\to Hi+2(X,Z)$, induces an isomorphism

$$L^j : H^{n-j}(X,\mathbb{Q}) \to H^{n+j}(X,\mathbb{Q}), \quad \forall j \leq n.$$

Not only the theorems of Lefschetz play an important role for our particular purposes, but we feel that we should also spend a few words sketching how they lead to some very interesting and still widely open conjectures, the Hartshorne conjectures (see [210]).

Assume now that the smooth projective variety $X\subset P^N$ is the complete intersection of $N-n$ hypersurfaces (this means the the sheaf IX of ideals of functions vanishing on X is generated by polynomials $F1,\ldots,Fc$, $c{:}=N-n$

being the codimension of X). Then the theorems of Lefschetz2 imply that the homology groups of Xequal those of P^N for $i \le n-1$ (recall that $Hi(P^N,Z)=0$ for i odd, while $Hi(P^N,Z)=Z$ for $i \le 2N$, i even). Similarly holds true for the homotopy groups, and we recall that, since $P^N=S^{2N+1}/S1$, then $\pi i(PN)=0$ for $i \le 2N, i \ne 0,2$.

It was an interesting discovery by Barth (see [22–24, 174] for the following and related results) that a similar (but weaker) result holds true for each smooth subvariety of PN, provided the codimension $c=N-n$ of X is smaller than the dimension.

Theorem 4 (Barth–Larsen) Let X be a smooth subvariety of dimension n in PN: then the homomorphisms

$$H_i(X, \mathbb{Z}) \to H_i(\mathbb{P}^N, \mathbb{Z}), \quad \pi_i(X) \to \pi_i(\mathbb{P}^N)$$

are bijective for $i \le n-c \Leftrightarrow i<2n-N+1$, and surjective for $i=n-c+1=2n-N+1$.

Observe that, if $N=n+1$, then the above result yields exactly the one of Lefschetz, hence the theorem is sharp in this trivial case (but much weaker for complete intersections of higher codimension). The case of the Segre embedding $X:=P1\times P2\to P5$ is a case which shows how the theorem is sharp since $(Z)^2 \cong H2(X,Z) \to H2(P^8,Z) \cong Z$ is surjective but not bijective.

The reader might wonder why the theorem of Barth and Larsen is a generalization of the theorem of Lefschetz. First of all, while Barth used originally methods of holomorphic convexity in complex analysis (somehow reminiscent of Morse theory in the real case) Hartshorne showed [210] how the third Lefschetz Theorem implies the result of Barth for cohomology with coefficients in Q. Moreover a strong similarity with the Lefschetz situation follows from the fact that one may view it (as shown by Badescu, see [18]) as an application of the classical Lefschetz theorem to the intersection

$$(X \times \mathbb{P}^N) \cap \Delta \cong X,$$

where $\Delta \subset P^N \times P^N$ is the diagonal. In turn, an idea of Deligne [133, 175] shows that the diagonal $\Delta \subset P^N \times P^N$ behaves 'like' a complete intersection, essentially because, under the standard birational map $PN\times PN \dashrightarrow P2N$, it maps to a linear subspace of P^{2N}.

The philosophy is then that smooth subvarieties of small codimension behave like complete intersections. This could be no accident if the well known Hartshorne conjecture [210] were true.

Conjecture 5 (On subvarieties of small codimension, Hartshorne) Let X be a smooth subvariety of dimension n in PN, and assume that the dimension is bigger than twice the codimension, $n>2(N-n)$: then X is a complete intersection.

While the conjecture says nothing in the case of curves and surfaces and is trivial in the case where $n\leq 4$, since a codimension 1 subvariety is defined by a single equation, it starts to have meaning for $n\geq 5$ and $c=N-n\geq 2$.

In the case where $c=N-n=2$, then by a result of Serre one knows (see [154], p. 143, also for a general survey of the Hartshorne conjecture for codimension 2 subvarieties) that X is the zero set of a section s of a rank 2 holomorphic vector bundle V on PN (observe that in this codimension one has that Pic(X)≅Pic(P^N), so that X satisfies the condition of being subcanonical: this means that $\omega X=O_X(d)$ for some d): in the case $c=2$the conjecture by Hartshorne is then equivalent to the conjecture

Conjecture 6 (On vector bundles of rank 2 on projective space, Hartshorne) Let V be a rank 2 vector bundle on PN, and assume that $N\geq 7$: then V is a direct sum of line bundles.

The major evidence for the conjecture on subvarieties of small codimension comes from the concept of positivity of vector bundles ([175], also [267]). In fact, many construction methods of subvarieties X of small codimension involve a realization of X as the locus where a vector bundle map drops rank to an integer r, and$X$ becomes singular if there are points of X in the locus Σ where the rank drops further down to $(r-1)$.

The expected dimension of Σ is positive in the range of Hartshorne's conjecture, but nevertheless this is not sufficient to show that $\Sigma\cap X$ is non empty.

Hartshorne's Conjecture 5 is related to projections: in fact, for each projective variety $X\subset$PN there exists a linear projection PN⇢P$2n+1$ whose restriction to X yields an embedding $X\subset$P$2n+1$: in other words, an embedding where the codimension is equal to the dimenion n plus 1. The condition of being embedded as a subvariety where the codimension is smaller or equal than the dimension is already a restriction (for instance, not all curves are plane curves, and smooth surfaces in P3 are simply connected by Lefschetz's theorem), and the smaller the codimension gets, the stronger the restrictions are (as shown by Theorem 4).

Speaking now in more technical terms, a necessary condition for *X* to be a complete intersection is that the sheaf of ideals I*X* be arithmetically Cohen-Macaulay (ACM, for short), which means that all higher cohomology groups *Hi*(P*N*,I*X*(*d*))=0 vanish for *n*≥*i*>0 and ∀*d*∈Z. In view of the exact sequence

$$0 \to \mathcal{I}_X \to \mathcal{O}_{\mathbb{P}^N} \to \mathcal{O}_X \to 0$$

the CM condition amounts to two conditions:

(1)*X* is projectively normal, i.e., the linear system cut on *X* by polynomials of degree *d* is complete (*H*⁰(P*N*,OP*N*(*d*))→*H*⁰(O*X*(*d*)) is surjective for all *d*≥0)

(2)*Hi*(*X*,O*X*(*d*))=0 for all *d*∈Z, and for all 0<*i*<*n*=dim(*X*).

In the case where *X* has codimension 2, and *N*≥6 (see [154], cor. 4.2, p. 165), the condition of being a complete intersection is equivalent to projective normality..

Linear normality is the case *d*=1 and means that *X* is not obtained as the projection of a non-degenerate variety from a higher dimensional projective space P*N*+1. This part of Hartshorne's conjecture is the only one which has been verified: a smooth subvariety with *n*≥2*c*−1=2(*N*−*n*)−1 is linearly normal (theorem of Zak [378]).

For higher *d*, one considers the so-called formal neighbourhood of *X*: denoting by $N_X^\vee := \mathcal{I}_X/\mathcal{I}_X^2$ the conormal bundle of *X*, one sees that, in order to show projective normality, in view of the exact sequence

$$0 \to \mathcal{I}_X^{m+1}(d) \to \mathcal{I}_X^m(d) \to Sym^m\left(N_X^\vee\right)(d) \cong \left(\mathcal{I}_X^m/\mathcal{I}_X^{m+1}\right)(d) \to 0,$$

a crucial role is played by the cohomology groups

$$H^q(Sym^m(N_X^\vee)(d))).$$

We refer the reader to [323] for a discussion of more general Nakano type vanishing statements of the form *Hq*(*Symm*(*N*∨*X*)(*d*)⊗Ω*pX*) (these could be implied by very strong curvature properties on the normal bundle, see also [267]).

Observe finally that the theorems of Lefschetz have been also extended to the case of singular varieties (see [172, 184]), but we shall not need to refer to these extensions in the present paper.

3 ALGEBRAIC TOPOLOGY: NON EXISTENCE AND EXISTENCE OF CONTINUOUS MAPS

The first famous achievements of algebraic topology were based on functoriality, which was used to infer the nonexistence of certain continuous maps.

The Brouwer's fixed point theorem says that every continuous self map $f:D^n \to D^n$, where $D^n=\{x \in R^n | |x| \leq 1\}$ is the unit disk, has a fixed point. The argument is by contradiction: otherwise, letting $\phi(x)$ be the intersection of the boundary $Sn-1$ of Dn with the half line stemming from $f(x)$ in the direction of x, ϕwould be a continuous map

$$\phi : D^n \to S^{n-1}, \quad s.t. \, \phi|_{S^{n-1}} = \mathrm{Id}_{S^{n-1}} .$$

The key point is to show that the reduced3 homology group $H_{n-1}(S^{n-1},Z) \cong Z$, while $Hn-1(Dn,Z)=0$, the disc being contractible; after that, denoting by $\iota:Sn-1 \to Dn$ the inclusion, functoriality of homology groups, since $\phi \circ \iota = \mathrm{Id}Sn-1$, would imply $0 = H_{n-1}(\phi) \circ H_{n-1}(\iota) = H_{n-1}(\mathrm{Id}_{S^{n-1}}) = \mathrm{Id}_{\mathbb{Z}}$, the desired contradiction.

Also well known is the Borsuk–Ulam theorem, asserting that there is no odd continuous function $F : S^n \to S^m$ for $n > m$ (odd means that $F(-x)=-F(x)$).

Here there are two ingredients, the main one being the cohomology algebra, and its contravariant functoriality: to any continuous map $f:X \to Y$ there corresponds an algebra homomorphism

$$f^* : H^*(Y, R) = \oplus_{i=0}^{\dim(Y)} H^i(Y, R) \to H^*(X, R),$$

for any ring R of coefficients.

In our case one takes as $X := \mathbb{P}^n_{\mathbb{R}} = S^n/\{\pm 1\}$, similarly $Y := \mathbb{P}^m_{\mathbb{R}} = S^m/\{\pm 1\}$ and lets f be the continuus map induced by F. One needs to show that, choosing $R=Z/2Z$, then the cohomology algebra of real projective space is a truncated polynomial algebra, namely:

$$H^*(\mathbb{P}^n_{\mathbb{R}}, \mathbb{Z}/2\mathbb{Z}) \cong (\mathbb{Z}/2\mathbb{Z})[\xi_n]/(\xi_n^{n+1}).$$

The other ingredient consists in showing that

$$f^*([\xi_m]) = [\xi_n],$$

[*ξm*] denoting the residue class in the quotient algebra.

One gets then the desired contradiction since, if *n*>*m*,

$$0 = f^*(0) = f^*([\xi_m]^{m+1}) = f^*([\xi_m])^{m+1} = [\xi_n]^{m+1} \neq 0.$$

Notice that up to now we have mainly used that *f* is a continuous map *f*:=PnR→PmR, while precisely in order to obtain that *f*∗([*ξm*])=[*ξn*] we must make use of the hypothesis that *f* is induced by an odd function *F*.

This property can be interpreted as the property that one has a commutative diagram

$$\begin{array}{ccc} S^n & \rightarrow & S^m \\ \downarrow & & \downarrow \\ \mathbb{P}^n_{\mathbb{R}} & \rightarrow & \mathbb{P}^m_{\mathbb{R}} \end{array}$$

which exhibits the two sheeted covering of $\mathbb{P}^n_{\mathbb{R}}$ by *Sn* as the pull-back of the analogous two sheeted cover for $\mathbb{P}^m_{\mathbb{R}}$. Now, as we shall digress soon, any such two sheeted covering is given by a homomorphism of *H*1(*X*,Z/2Z)→Z/2Z, i.e., by an element in *H*1(*X*,Z/2Z), and this element is trivial if and only if the covering is trivial (that is, homeomorphic to *X*×(Z/2Z), in other words a disconnected cover).

This shows that the pull back of the cover, which is nontrivial, corresponds to *f*∗([*ξm*]) and is nontrivial, hence *f*∗([*ξm*])=[*ξn*].

As we saw already in the first section, algebraic topology attaches to a good topological space homology groups *Hi*(*X*,*R*), which are covariantly functorial, a cohomology algebra *H*∗(*X*,*R*) which is contravariantly functorial, and these groups can be calculated, by virtue of the Mayer Vietoris exact sequence and of excision (see any textbook), by chopping the space in smaller pieces. In particular, these groups vanish when *i*>*dim*(*X*). But to *X* are also attached the homotopy groups *πi*(*X*). The common feature is that homotopic maps induce the same homomorphisms on homology, cohomology, and homotopy.

We are, for our purposes, more interested in the more mysterious homotopy groups, which, while not necessarily vanishing for $i>dim(X)$, enjoy however a fundamental property.

Recall the definition due to Whitney and Steenrod [350] of a fibre bundle. In the words of Steenrod, the notion of a fibre bundle is a weakening of the notion of a product, since a product $X\times Y$ has two continuous projections $pX:X\times Y\to X$, and $pY:X\times Y\to Y$, while a fibre bundle E over B with fibre F has only one projection, $p=pB:E\to B$ and its similarity to a product lies in the fact that for each point $x\in B$ there is an open set U containing x, and a homeomorphism $p-1B(U)\cong U\times F$ compatible with both projections onto U.

The fundamental property of fibre bundles is that there is a long exact sequence of homotopy groups

$$\cdots \to \pi_i(F) \to \pi_i(E) \to \pi_i(B) \to \pi_{i-1}(F) \to \pi_{i-1}(E) \to \pi_{i-1}(B) \to \cdots$$

where one should observe that $\pi i(X)$ is a group for $i\geq 1$, an abelian group for $i\geq 2$, and for $i=0$ is just the set of arc-connected components of X (we assume the spaces to be good, that is, locally arcwise connected, semilocally simply connected, see [189], and, most of the times, connected).

The special case where the fibre F has the discrete topology is the case of a **covering space**, which is called the universal covering if moreover $\pi 1(E)$ is trivial.

Special mention deserves the following more special case.

Definition 7 Assume that E is arcwise connected, contractible (hence all homotopy groups $\pi i(E)$ are trivial), and that the fibre F is discrete, so that all the higher homotopy groups $\pi i(B)=0$ for $i\geq 2$, while $\pi 1(B)\cong\pi 0(F)=F$. Then one says that B is a classifying space $K(\pi,1)$ for the group $\pi=\pi 1(B)$.

In general, given a group π, a CW complex B is said to be a $K(\pi,1)$ if $\pi i(B)=0$ for $i\geq 2$, while $\pi 1(B)\cong\pi$.

Example 8 The easiest examples are the following ones, where (2) is a case where we have a complex projective variety (see Sect. 3 for more such examples):

(1) the real torus $T^n:=R^n/Z^n$ is a classifying space $K(Z^n,1)$ for the group $\pi=Z^n$;

*(2)*a complex projective curve C of genus $g\geq 2$ is a classifying space $K(\pi g,1)$, since by the uniformization theorem its universal covering is the Poincaré upper half plane H:={z∈C|$Im(z)$>0} and its fundamental group $\pi 1(C)$ is isomorphic to the group

$$\pi_g := \langle \alpha_1, \beta_1, \dots \alpha_g, \beta_g | \Pi_1^g [\alpha_i, \beta_i] = 1 \rangle,$$

quotient of a free group with $2g$ generators by the normal subgroup generated by the relation $\Pi_1^g [\alpha_i, \beta_i];$;

*(3)*a classifying space K(Z/2Z,1) is given by the inductive limit $\mathbb{P}^{\infty}_{\mathbb{R}} := \lim_{n\to\infty} \mathbb{P}^n_{\mathbb{R}}$. To show this, it suffices to show that $S\infty$:=lim$n\to\infty Sn$ is contractible or, equivalently, that the identity map is homotopic to a constant map. We do this as follows:4 let σ:R∞→R∞ be the shift operator, and first define a homotopy of the identity map of R∞\{0} to the constant map with value $e1$. The needed homotopy is the composition of two homotopies:

$$F(t, v) := (1-t)v + t\sigma(v), \quad 0 \leq t \leq 1,$$
$$F(t, v) := (2-t)\sigma(v) + (t-1)e_1, \quad 1 \leq t \leq 2, \quad \forall v \in \mathbb{R}^{\infty}.$$

Then we simply project the homotopy from $\mathbb{R}^{\infty}\backslash\{0\}$ to S^{∞} considering $\frac{F(t,v)}{|F(t,v)|}$.

These classifying spaces, although not unique, are unique up to **homotopy-equivalence** (we use the notation $X\sim h.e.Y$ to denote homotopy equivalence: this means that there exist continuous maps f:$X\to Y$, g:$Y\to X$ such that both compositions $f\circ g$ and $g\circ f$ are homotopic to the identity).

Therefore, given two classifying spaces for the same group, they not only do have the same homotopy groups, but also the same homology and cohomology groups. Thus the following definition is well posed.

Definition 9 Let Γ be a finitely presented group, and let $B\Gamma$ be a classifying space for Γ: then the homology and cohomology groups and algebra of Γ are defined as

$$H_i(\Gamma, \mathbb{Z}) := H_i(B\Gamma, \mathbb{Z}),\ H^i(\Gamma, \mathbb{Z}) := H^i(B\Gamma, \mathbb{Z}),\ H^*(\Gamma, \mathbb{Z}) := H^*(B\Gamma, \mathbb{Z}),$$

and similarly for other rings of coefficients instead of Z.

Remark 10 The concept of a classifying space *BG* is indeed more general: the group *G* could also be a Lie group, and then, if *EG* is a contractible space over which *G* has a free (continuous) action, then one defines *BG*:=*EG*/*G*.

The typical example is the simplest compact Lie group *G*=*S*1: then, keeping in mind that *S*1={*z*∈C||*z*|=1}, we take as *EG* the space

$$ES^1 := S(\mathbb{C}^\infty) = lim_{n\to\infty} S(\mathbb{C}^n) = lim_{n\to\infty} \{w \in \mathbb{C}^n \| w| = 1\}.$$

Classifying spaces, even if often quite difficult to construct explicitly, are very important because they guarantee the existence of continuous maps! We have more precisely the following (cf. [349], Theorem 9, p. 427, and Theorem 11, p. 428)

Theorem 11

Let *Y* be a 'nice' topological space, i.e., *Y* is homotopy-equivalent to a CW-complex, and let *X* be a nice space which is a *K*(*π*,1) space: then, choosing base points *y*0∈*Y*,*x*0∈*X*, one has a bijective correspondence

$$[(Y, y_0), (X, x_0)] \cong Hom(\pi_1(Y, y_0), \pi_1(X, x_0)), \quad [f] \mapsto \pi_1(f),$$

where [(*Y*,*y*0),(*X*,*x*0)] denotes the set of homotopy classes [*f*] of continuous maps *f*:*Y*→*X* such that *f*(*y*0)=*x*0 (and where the homotopies *F*(*y*, *t*) are also required to satisfy *F*(*y*0,*t*)=*x*0, ∀*t*∈[0,1]).

In particular, the free homotopy classes [*Y*, *X*] of continuous maps are in bijective correspondence with the conjugacy classes of homomorphisms Hom(*π*1(*Y*,*y*0),*π*) (conjugation is here inner conjugation by Inn(*π*) on the target).

Observe that, quite generally, the universal covering *Eπ* of a classifying space *Bπ*:=*K*(*π*,1) associates (by the lifting property) to a continuous map *f*:*Y*→*Bπ* a *π*1(*Y*)-equivariant map ˜*f*

$$\tilde{f} : \tilde{Y} \to E_\pi,$$

where the action of *π*1(*Y*) on *Eπ* is determined by the homomorphism *φ*:= $\pi_1(Y) \to \pi = \pi_1(B\pi).$.

Moreover, any *φ*:*π*1(*Y*)→*π*=*π*1(*Bπ*) determines a fibre bundle *Eφ* over *Y* with fibre *Eπ*:

$$E_\varphi := (\tilde{Y} \times E_\pi)/\pi_1(Y) \to Y = (\tilde{Y})/\pi_1(Y) = Y,$$

where the action of $\gamma \in \pi 1(Y)$ is as follows: $\gamma(y',v)=(\gamma(y'),\varphi(\gamma)(v))$.

While topology deals with continuous maps, when dealing with manifolds more regularity is wished for. For instance, when we choose for Y a differentiable manifold M, and the group π is abelian and torsion free, say $\pi = Zr$, then a more precise incarnation of the above theorem is given by the De Rham theory.

In fact, a homomorphism $\varphi:\pi 1(Y)\to Zr$ factors through the Abelianization $H1(Y,Z)$ of the fundamental group. Since $H^1(Y,Z)=\mathrm{Hom}(H^1(Y,Z),Z)$, φ is equivalent to giving an element in

$$\varphi \in H^1(Y,\mathbb{Z})^r \subset H^1(Y,\mathbb{R})^r \cong H^1_{DR}(Y,\mathbb{R})^r,$$

where $H^1_{DR}(Y,\mathbb{R})$ is the quotient space of the space of closed differentiable 1-forms modulo exact 1-forms.

In this case the classifying space is a real torus

$$T_r := \mathbb{R}^r/\mathbb{Z}^r.$$

Observe however that to give $\varphi:\pi 1(Y)\to Zr$ it is equivalent to give its r components $\varphi i, i=1,\dots,r$, which are homomorphisms into Z, and giving a map to $Tr{:=}Rr/Zr$ is equivalent to giving r maps to $T1{:=}R/Z$: hence we may restrict ourselves to consider the case $r=1$.

Let us sketch the basic idea of the previous Theorem 11 in this special case. Let us assume that Y is a cell complex, and define as usual Yj to be its j-th skeleton, the union of all the cells of dimension $i \le j$.

Since the fundamental group of Y is generated by the free group $F{:=}\pi 1(Y1)$, we get a homomorphism $\Phi:F\to Z$ inducing φ. For each 1-cell $\gamma \cong S1$ we send $\gamma \to S1$ according to the map $z \in S1 \mapsto zm \in S1$, where $m=\Phi(\gamma)$.

In this way we get a continuous map $f1:Y1\to S1$, and we want to extend it inductively to Yj for each j. Now, assume that f is already defined on Z, and that you are attaching an n-cell to Z, according to a continuous map $\psi:\partial(Dn)=Sn-1\to Z$. In order to extend f to $Z \cup \psi Dn$ it suffices to extend the map $f \circ \psi$ to the interior of the disk Dn. This is possible once the map $f \circ \psi : Sn-1 \to S1$ is homotopic to a constant map. Now, for $n=2$, this condition holds by assumption: since $\psi(S1)$ yields a relation for $\pi 1(Y)$, therefore its image under ϕ must be equal to zero.

For higher n, $n\geq 3$, it suffices to observe that a continuous map $h:Sn-1\to S1$ extends to the interior always: since $Sn-1$ is simply connected, and $S1=R/Z$, h lifts to a continuous map $h':Sn-1\to R$, and we can extend h' to Dn by setting

$$h'(x) := |x|h'\left(\frac{x}{|x|}\right), \quad \forall x \neq 0, \quad h'(0) := 0.$$

(h is then the composition of h' with the projection $R\to R/Z=S1$).

In general, when both Y and the classifying space X (as $T1$ here) are differentiable manifolds, then each continuous map $f:Y\to X$ is homotopic to a differentiable map f'. Take in fact $X\subset R^N$ and observe that the implicit function theorem implies that there is a tubular neighbourhood $X\subset TX\subset R^N$ diffeomorphic to a tubular neighbourhood of X embedded as the 0-section of the normal bundle NX of the embedding $X\subset R^N$.

Therefore, approximating the function $f:Y\to RN$ by a differentiable function f'' with values in TX, we can use the bundle projection $NX\to X$ to project f'' to a differentiable function $f':Y\to X$, and similarly we can project the natural homotopy between f and f'', $f(y)+tf''(y)$ to obtain a homotopy between f and f'.

Once we have a differentiable map $f':Y\to T1=R/Z$, we simply take the lift $\sim f':\tilde{}Y\to R$, and the differential $d\sim f'$ descends to a closed differential form η on Y such that its integral over a closed loop γ is just $\phi(\gamma)$.

We obtain the

Proposition 12 Let Y be a differentiable manifold, and let X be a differentiable manifold that is a $K(\pi,1)$ space: then, choosing base points $y0\in Y, x0\in X$, one has a bijective correspondence

$$[(Y, y_0), (X, x_0)]^{diff} \cong Hom(\pi_1(Y), \pi), \quad [f] \mapsto \pi_1(f),$$

where $[(Y, y_0), (X, x_0)]^{diff}$ denotes the set of differential homotopy classes $[f]$ of differentiable maps $f:Y\to X$ such that $f(y0)=x0$.

In the case where X is a torus $Tr=R^r/Z^r$, then f is obtained as the projection onto Tr of

$$\tilde{\phi}(y) := \int_{y_0}^{y} (\eta_1, \ldots, \eta_r), \eta_j \in H^1(Y, \mathbb{Z}) \subset H^1_{DR}(Y, \mathbb{R}).$$

Remark 13 In the previous proposition, ηj is indeed a closed 1-form, representing a certain De Rham cohomology class with integral periods (i.e., $\int_{y_0}^{y}(\eta_1,\ldots,\eta_r)\ \mathrm{mod}(\mathbb{Z}^r).$). Therefore f is defined by $\int yy0(\eta 1,\ldots,\eta r)$ mod(Z*r*). Moreover, changing ηj with another form $\eta j+dFj$ in the same cohomology class, one finds a homotopic map, since $\int_{y_0}^{y}(\eta_j + t dF_j) = \int_{y_0}^{y}(\eta_j) + t(F_j(y) - F_j(y_0)).$

Before we dwell into a review of results concerning higher regularity of the classifying maps, we consider in the next section the basic examples of projective varieties that are classifying spaces.

4 PROJECTIVE VARIETIES WHICH ARE *K(Π,1)*

The following are the easiest examples of projective varieties which are $K(\pi,1)$'s.

(1) Projective curves C of genus $g\geq 2$.

By the **Uniformization theorem**, these have the Poincaré upper half plane H:={z∈C|$Im(z)$>0} as universal covering, hence they are compact quotients C=H/Γ, where Γ⊂PSL(2,R) is a discrete subgroup isomorphic to the fundamental group of C, $\pi 1(C)\cong\pi g$. Here

$$\pi_g := \langle \alpha_1, \beta_1, \ldots \alpha_g, \beta_g | \Pi_1^g [\alpha_i, \beta_i] = 1 \rangle$$

contains no elements of finite order. Hence, given a faithful action of πg on H, it follows that necessarily Γ acts freely on H. Moreover, the quotient must be compact, otherwise C would be homotopically equivalent to a bouquet of circles, hence $H2(C,$Z$)=0$, a contradiction, since $H2(C,$Z$)\cong H2(\pi g,$Z$)\cong$Z, as one sees taking the standard realization of a classifying space for πg by glueing the $2g$ sides of a polygon in the usual pattern.

Moreover, the complex orientation of C induces a standard generator [C] of $H2(C,$Z$)\cong$Z, the so-called fundamental class.

(2) AV : = Abelian varieties.

More generally, a complex torus X=Cg/Λ, where Λ is a discrete subgroup of maximal rank (isomorphic then to Z2g), is a Kähler classifying space K(Z2g,1), the Kähler metric being induced by the translation invariant Euclidean metric $i2\sum g1dzj\otimes d\bar{}zj$.

For g=1 one gets in this way all projective curves of genus g=1; but, for g>1, X is in general not projective: it is projective, and called then an Abelian variety, if it satisfies the Riemann bilinear relations. These amount to the existence of a positive definite Hermitian form H on Cg whose imaginary part A (i.e., H=S+iA), takes integer values on Λ×Λ. In modern terms, there exists a positive line bundle L on X, with Chern class A∈$H$2(X,Z)=$H$2(Λ,Z)=Λ2(Hom(Λ,Z)), whose curvature form, equal to H, is positive (the existence of a positive line bundle on a compact complex manifold X implies that X is projective algebraic, by Kodaira's theorem, [247]).

We shall indeed see (82) that Abelian varieties are exactly the projective $K(\pi,1)$ varieties, for which π is an abelian group.

(3) LSM : = Locally symmetric manifolds.

These are the quotients of a **bounded symmetric domain** D by a cocompact discrete subgroup Γ⊂Aut(D)acting freely. Recall that a bounded symmetric domain D is a bounded domain D⊂⊂Cn such that its group Aut(D) of biholomorphisms contains for each point p∈D, a holomorphic automorphism σp such that $\sigma p(p)=p$, and such that the derivative of σp at p is equal to $-Id$. This property implies that σ is an involution (i.e., it has order 2), and that Aut(D)0 (the connected component of the identity) is transitive on D, and one can write D=G/K, where G is a connected Lie group, and K is a maximal compact subgroup.

The two important properties are:

(3.1) D splits uniquely as the product of irreducible bounded symmetric domains.

(3.2) each such D is contractible, since there is a Lie subalgebra L of the Lie algebra G of G such that the exponential map is a homeomorphism L≅D. Hence X is a classifying space for the group Γ≅π1(X).

Bounded symmetric domains were classified by Elie Cartan [77], and there is only a finite number of them (up to isomorphism) for each dimension n.

Recall the notation for the irreducible domains:

(i)*In,p* is the domain D={Z∈Mat(n,p,C):Ip−tZ·¯Z>0}.

(ii)*IIn* is the intersection of the domain *In,n* with the subspace of skew symmetric matrices.

(iii)*IIIn* is instead the intersection of the domain *In,n* with the subspace of symmetric matrices.

(iv)The Cartan–Harish Chandra realization of a domain of type *IVn* in C*n* is the subset D defined by the inequalities (compare [213], p. 527)

$$|z_1^2 + z_2^2 + \cdots + z_n^2| < 1,$$
$$1 + |z_1^2 + z_2^2 + \cdots + z_n^2|^2 - 2\left(|z_1|^2 + |z_2|^2 + \cdots + |z_n|^2\right) > 0.$$

(v)D16 is the exceptional domain of dimension *d*=16.

(vi)D27 is the exceptional domain of dimension *d*=27.

We refer the reader to [213], Theorem 7.1, p. 383 and exercise D, pp. 526–527, and [329] p. 525 for a list of these irreducible bounded symmetric domains, and a description of all of them as homogeneous spaces*G* / *K*. In this context the domains are also called **Hermitian symmetric spaces of non compact type**. Each of these is contained in the so-called **compact dual**, which is a Hermitian symmetric spaces of compact type.

The easiest example is, for type I, the Grassmann manifold. For type IV, the compact dual of D is the hyperquadric *Qn*⊂P*n*+1 defined by the polynomial $\sum_{j=0}^{n-1} X_j^2 - X_n^2 - X_{n+1}^2$. Notice that SO0(*n*,2)⊂Aut(*Qn*). The Borel embedding *j*:D→*Qn* is given by

$$j(z_1, \ldots, z_n) = [2z_1 : 2z_2 : \cdots : 2z_n : \mathrm{i}(\Lambda - 1) : \Lambda + 1],$$

where $\Lambda := z_1^2 + \cdots + z_n^2$. The map *j* identifies the domain D with the SO0(*n*,2)-orbit of the point [0:0:⋯:1:i]∈*Qn*, i.e. D≅SO0(*n*,2)/ SO(*n*)× SO(2).

Among the bounded symmetric domains are the so called **bounded symmetric domains of tube type**, those which are biholomorphic to a **tube domain**, a generalized Siegel upper half-space

$$T_C = \mathbb{V} \oplus \sqrt{-1}C$$

where V is a real vector space and C⊂V is a *symmetric cone*, i.e., a self dual homogeneous convex cone containing no full lines.

In the case of type III domains, the tube domain is Siegel's upper half space:

$$\mathcal{H}_g := \left\{\tau \in \mathrm{Mat}(g, g, \mathbb{C}) | \tau = {}^t\tau, \mathrm{Im}(\tau) > 0\right\},$$

a generalisation of the upper half-plane of Poincaré.

Borel proved in [57] that for each bounded symmetric domain D there exists a compact free quotient X=D/Γ, called a compact Clifford–Klein form of the symmetric domain D.

A classical result of Hano (see [207] Theorem IV, p. 886, and Lemma 6.2, p. 317 of [292]) asserts that a bounded homogeneous domain that is the universal cover of a compact complex manifold is symmetric.

(4) A particular, but very explicit case of locally symmetric manifolds is given by the VIP : = Varieties isogenous to a product.

These were studied in [96], and they are defined as quotients

$$X = (C_1 \times C_2 \times \cdots \times C_n)/G$$

of the product of projective curves Cj of respective genera $gj \geq 2$ by the action of a finite group G acting freely on the product.

In this case the fundamental group of X is not so mysterious and fits into an exact sequence

$$1 \to \pi_1(C_1 \times C_2 \times \cdots \times C_n) \cong \pi_{g_1} \times \cdots \times \pi_{g_n} \to \pi_1(X) \to G \to 1.$$

Such varieties are said to be of the **unmixed type** if the group G does not permute the factors, i.e., there are actions of G on each curve such that

$$\gamma(x_1, \ldots, x_n) = (\gamma x_1, \ldots, \gamma x_n), \forall \gamma \in G.$$

Equivalently, each individual subgroup πgj is normal in $\pi 1(X)$.

(5) Kodaira fibrations $f: S \to B$.

Here S is a smooth projective surface and all the fibres of f are smooth curves of genus $g \geq 2$, in particular f is a differentiable fibre bundle. Unlike the examples above, where $C=(C1 \times C2)/G \to C2/G$ is a holomorphic fibre bundle with fibre $C1$ if G acts freely on $C2$, the second defining property for Kodaira fibrations is that the fibres are not all biholomorphic to each other.

These Kodaira fibred surfaces S are very interesting topological objects: they were constructed by Kodaira [251] as a counterexample to the conjecture that the index (of the cup product in middle cohomology) would be multiplicative for fibre bundles. In fact, for curves the pairing $H^1(C,\mathbb{Z})\times H^1(C,\mathbb{Z})\to H^2(C,\mathbb{Z})\cong\mathbb{Z}$ is skew symmetric, hence it has index zero; while Kodaira showed that the index of the cup product $H^2(S,\mathbb{Z})\times H^2(S,\mathbb{Z})\to H^4(S,\mathbb{Z})\cong\mathbb{Z}$ is strictly positive.

The fundamental group of S fits obviously into an exact sequence:

$$1 \to \pi_g \to \pi_1(S) \to \pi_b \to 1,$$

where g is the fibre genus and b is the genus of the base curve B, and it is known that $b\geq 2$, $g\geq 3$ (see [109], also for more constructions and a thorough discussion of their moduli spaces).

By simultaneous uniformization ([40]) the universal covering $\tilde{S}$ of a Kodaira fibred surface S is biholomorphic to a bounded domain in $\mathbb{C}^2$ (fibred over the unit disk $\Delta:=\{z\in\mathbb{C}\,|\,|z|<1\}$ with fibres isomorphic to Δ), which is not homogeneous.

(6) Hyperelliptic surfaces: these are the quotients of a complex torus of dimension 2 by a finite group G acting freely, and in such away that the quotient is not again a complex torus.

These surfaces were classified by Bagnera and de Franchis ([19], see also [21, 156]) and they are obtained as quotients $(E_1\times E_2)/G$ where E_1,E_2 are two elliptic curves, and G is an abelian group acting on E_1 by translations, and on E_2 effectively and in such a way that $E_2/G\cong\mathbb{P}^1$.

(7) In higher dimension we define the Generalized Hyperelliptic Varieties (GHV) as quotients $A\,/\,G$ of an Abelian Variety A by a finite group G acting freely, and with the property that G is not a subgroup of the group of translations. Without loss of generality one can then assume that G contains no translations, since the subgroup G_T of translations in G would be a normal subgroup, and if we denote $G'=G/G_T$, then $A/G=A'/G'$, where A' is the Abelian variety $A':=A/G_T$.

We propose instead the name **Bagnera–de Franchis (BdF) Varieties** for those quotients $X=A/G$ were Gcontains no translations, and G is a cyclic group of order m, with generator g (observe that, when A has dimension $n=2$, the two notions coincide, thanks to the classification result of Bagnera and De Franchis [19]).

A concrete description of such Bagnera–De Franchis varieties shall be given in the following section.

We end this section giving an example of a projective variety which is not a $K(\pi,1)$, thus showing that the property of being a projective classifying space is lost after taking hyperplane sections.

Proposition 14 Let $n \geq 3$, and consider $X = C1 \times \cdots \times Cn$, the product of n projective curves Ci of respective genera $gi \geq 1$. Let $X \subset PN$ be a projective embedding, and let S be a smooth surface, obtained taking the complete intersection of X with $n-2$ hypersurfaces. Then $\pi 2(S) \neq 0$, in particular, S is not a projective classifying space.

Proof By the theorem of Lefschetz $\pi 1(S) \cong \pi 1(X)$, hence the universal covering ˜S of S is a closed complex submanifold of ˜$X = \mathrm{C}r \times \mathrm{H}n-r$. Hence ˜$S$ is a Stein manifold, therefore (see [4]) it has the homotopy type of a CW complex of real dimension ≤2.

Since $\pi 1(\tilde{}S) = 0$, we claim that $H2(\tilde{}S, \mathrm{Z}) = \pi 2(\tilde{}S) = \pi 2(S) \neq 0$ (the first isomorphism follows from the theorem of Hurewicz).

Otherwise, all homology and homotopy groups of ˜S would be trivial and ˜S would be contractible, S would be a$K(\pi,1)$, hence $H*(S,\mathrm{Z}) = H*(\pi 1(S),\mathrm{Z}) = H*(X,\mathrm{Z})$. This is a contradiction, since $H2n(X,\mathrm{Z}) \neq 0$, $H2n(S,\mathrm{Z}) = 0$, by our hypothesis that $n \geq 3$. ▯

5 A TRIP AROUND BAGNERA–DE FRANCHIS VARIETIES AND GROUP ACTIONS ON ABELIAN VARIETIES

5.1 Bagnera–de Franchis varieties

Let A / G be a Generalized Hyperelliptic Variety. An easy observation is that any $g \in G$ is induced by an affine transformation $x \mapsto \alpha x + b$ on the universal cover, hence it does not have a fixed point on $A = V/\Lambda$ if and only if there is no solution of the equation

$$\exists x \in V,\ g(x) \equiv x \ (\bmod \Lambda)$$
$$\Leftrightarrow \exists x \in V, \lambda \in \Lambda,\ (\alpha - \mathrm{Id})x = \lambda - b.$$

This remark implies that 1 must be an eigenvalue of α for all non trivial transformations $g \in G$.

Since $\alpha \in GL(\Lambda) \cong GL(2n,\mathbb{Z})$, the eigenspace $V1 = Ker(\alpha - \mathrm{Id})$ is a complex subspace defined over Q, hence we have an Abelian subvariety

$$A_1 \subset A, A_1 := V_1/\Lambda_1, \quad \Lambda_1 := (V_1 \cap \Lambda).$$

While what we said up to now was valid for any complex torus, we replace this assumption by the stronger assumption that A is an Abelian variety. This assumption allows us to invoke Poincaré' s complete reducibility theorem, so that we can split

$$V = V_1 \oplus V_2, \; V_2 := (V_1)^{\perp}, \quad A_2 := V_2/\Lambda_2, \quad \Lambda_2 := (V_2 \cap \Lambda).$$

We get then an isogeny $A_1 \times A_2 \to A$, with kernel $T := \Lambda/(\Lambda_1 \oplus \Lambda_2)$. Observe that $T \cap Aj = \{0\}, j = 1,2$.

In view of the splitting $V = V1 \oplus V_2$, we can write, after a change of the origin in the affine space V_2,

$$\alpha(v_1 + v_2) = v_1 + \alpha_2 v_2, g(v) = v_1 + b_1 + \alpha_2 v_2, \quad b_1 \in V_1, b_1 \notin \Lambda.$$

If $\alpha 2$ has order equal to m, then necessarily the image of $b1$ has order exactly m in A, by virtue of our assumption that G contains no translations. In other words, $b1$ induces a translation on $A1$ of order exactly m.

Now, g lifts naturally to $A1 \times A2$, by

$g(a1,a2) = (a1 + [b1], \alpha 2 a2)$,

where $[b1]$ is the class of $b1$ in $A1$.

We reach the conclusion that $X = A/G = ((A_1 \times A_2)/T)/G$, where T is the finite group of translations $T = \Lambda/(\Lambda_1 \oplus \Lambda_2)$.

Conversely, given such an automorphism g of $A_1 \times A_2$, it descends to $A := (A_1 \times A_2)/T$ if and only if the linear part of g sends T to T.

Denote now by Tj the (isomorphic) image of $T \to Aj$: then $T \subset T_1 \times T_2$ is the graph of an isomorphism $\phi: T_1 \to T_2$, hence the condition which allows g to descend to A is that:

$$(**)(\mathrm{Id} \times \alpha_2)(T) = T \Leftrightarrow \alpha_2 \circ \phi = \phi \Leftrightarrow$$
$$\Leftrightarrow (\alpha_2 - \mathrm{Id}) \circ \phi = 0 \Leftrightarrow (\alpha_2 - \mathrm{Id})T_2 = 0.$$

We are in the position to illustrate the standard example, before we give the more general more complete description of BdF varieties.

The basic example is the one where m=2, hence $\alpha 2$ is scalar multiplication by −1. Then ϕ=−ϕ implies that T_2,T_1 are 2-torsion subgroups. Then also 2[b1]=0 implies that [b1] is a 2-torsion element. However [b1]cannot belong to $T$1, else g:A→A would be induced by

(a1,a2)↦(a1+[b1],α2a2)≡(a1,α2a2−ϕ([b_1])) (*mod T*),

which has a fixed point on A.

We conclude that the standard example of Bagnera–de Franchis Varieties of order m=2 is the following:

$$\begin{aligned} X &= A/g, \quad \text{where } A := (A_1 \times A_2)/T, \quad \text{and} \\ T &= \{(t_1, \phi(t_1))\}, T_j \subset A_j[2], \phi : T_1 \cong T_2, \beta_1 \in A_1[2] \backslash T_1, \\ g : A \to A, & \quad g(a_1, a_2) = (a_1 + \beta_1, -a_2). \end{aligned}$$

In order to conclude appropriately the above discussion, we give some useful definition.

Definition 15 We define first a Bagnera–de Franchis manifold (resp.: variety) **of product type** as a quotient X=A/G where$A$=$A$1×$A$2, $A$1,$A$2 are complex tori (resp.: Abelian Varieties), and G≅Z/m is a cyclic group operating freely on A, generated by an automorphism of the form

g(a1,a2)=(a1+β1,α2(a2)),

where β1∈$A$1[m] is an element of order exactly m, and similarly α2:$A$2→$A$2 is a linear automorphism of order exactly m without 1 as eigenvalue (these conditions guarantee that the action is free). If moreover all eigenvalues of α2 are primitive m-th roots of 1, we shall say that X=A/G is a **primary** Bagnera–de Franchis manifold.

We have the following proposition, giving a characterization of Bagnera-De Franchis varieties.

Proposition 16 Every Bagnera–de Franchis variety $X=A/G$, where $G\cong\mathbb{Z}/m$ contains no translations, is the quotient of a Bagnera–de Franchis variety of product type, $(A1\times A2)/G$ by any finite subgroup T of $A1\times A2$ which satisfies the following properties:

(1) T is the graph of an isomorphism between two respective subgroups $T1\subset A1, T2\subset A2$,

(2) $(\alpha 2-\mathrm{Id})T2=0$

(3) if $g(a_1,a_2)=(a_1+\beta_1,\alpha_2(a_2))$, then the subgroup of order m generated by $\beta 1$ intersects $T1$ only in $\{0\}$.

In particular, we may write X as the quotient $X=(A_1\times A_2)/(G\times T)$ by the abelian group $G\times T$.

5.2 Actions of a finite group on an Abelian variety

Assume that we have the action of a finite group G on a complex torus $A=V/\Lambda$. Since every holomorphic map between complex tori lifts to a complex affine map of the respective universal covers, we can attach to the group G the group of affine transformations Γ which fits into the exact sequence:

$0\to\Lambda\to\Gamma\to G\to 1.$

Γ consists of all affine transformations of V which lift transformations of the group G.

Define now G^0 to be the subgroup of G consisting of all the translations in G.

Proposition 17 The abstract group Γ determines an exact sequence

$0\to\Lambda'\to\Gamma\to G'\to 1$

such that $\Lambda'\subset\Lambda$, Λ' is a lattice in V, $\Lambda'/\Lambda=G0$, $G'\subset Aut(V/\Lambda')$ contains no translations.

Proof It is clear that $V=\Lambda\otimes ZR$ as a real vector space, and we denote by $VQ:=\Lambda\otimes Q$. Consider the homomorphism αL associating to an affine transformation its linear part, and let

$$\Lambda' := \ker(\alpha_L : \Gamma \to \mathrm{GL}(V_{\mathbb{Q}}) \subset \mathrm{GL}(V)),$$
$$\overline{G}_1 := \mathrm{Im}(\alpha_L : \Gamma \to \mathrm{GL}(V_{\mathbb{Q}})).$$

Λ' is a subgroup of the group of translations in V, hence it is obviously Abelian, and maps isomorphically onto a lattice of V which contains Λ. We shall identify this lattice with Λ', writing with shorthand notation $\Lambda'\subset V$.

In turn $V=\Lambda'\otimes ZR$, and, if $G':=\Gamma/\Lambda'$, then $G'\cong{}^{-}G1$ and we have the exact sequence

$0\to\Lambda'\to\Gamma\to G'\to 1$,

yielding an embedding $G'\subset GL(\Lambda')$.

There remains to show that Λ' is determined by Γ as an abstract group, independently of the exact sequence we started with. In fact, one property of Λ' is that it is a maximal abelian subgroup, normal and of finite index.

Assume that Λ'' has the same property: then their intersection $\Lambda 0:=\Lambda'\cap\Lambda''$ is a normal subgroup of finite index, in particular $\Lambda 0\otimes ZR=\Lambda'\otimes ZR=V$; hence $\Lambda''\subset\ker(\alpha L{:}\Gamma\to \mathrm{GL}(V))=\Lambda'$, where αL is induced by conjugation on $\Lambda 0$ By maximality $\Lambda'=\Lambda''$. ?

In the case where A is an Abelian variety, we can say a little bit more about the affine representation of the group Γ.

Lemma 18 Let A an Abelian variety, and let G be a finite group of automorphisms. Then there is a positive integer r such that the group Γ of affine transformations which are lifts of transformations in G satisfies

$$\Gamma \subset Aff\left(\frac{1}{r}\Lambda\right).$$

In other words, we can write any transformation $g\in G$ as induced by the affine transformation of V: $v\mapsto\alpha(v)+\lambda r,\lambda\in\Lambda$.

Proof Let L be an ample divisor class on A. Since G is finite, there is a G-invariant such class L (simply replace L by $\sum_{g \in G} g^*(L') =: L'$).

We can further assume that the class of L is not only G-invariant , but also indivisible. There is a homomorphism $\phi_L : A \to A^\vee$, such that $\phi_L(a) := T^*_a L - L \in Pic^0(A) = A^\vee$, and where T_a is the translation by a, $v \mapsto v + a$.

Let $K_L := \ker(\Phi_L)$, which is a finite group, and denote by r its exponent (hence, $rK_L = 0$ and $K_L \subset (\frac{1}{r}\Lambda)/\Lambda$).

Represent now L by a line bundle whose cocycle is in Appell-Humbert normal form $f_\lambda(z) = \rho(\lambda) exp(\pi H(z, \lambda) + \frac{\pi}{2} H(\lambda, \lambda))$, and write a transformation $g \in G$ as induced by $v \mapsto \alpha(v) + b$. The condition that the Chern class of L is G-invariant implies that the Hermitian form H on V is left invariant by α (i.e., we have a group of isometries of V for the metric given by the Hermitian form H).

Hence also the translation part leaves L invariant, therefore the class of b lies in $K_L \subset (\frac{1}{r}\Lambda)/\Lambda$.

Observe now that, given an affine group $\Gamma \subset Aff(\Lambda \otimes_{\mathbb{Z}} \mathbb{R})$ as above, in order to obtain the structure of a complex torus on V/Λ', we must give a complex structure on V which makes the action of $G' \cong \bar{G}_1$ complex linear.

In order to study the moduli spaces of the associated complex manifolds, we introduce therefore a further invariant, called Hodge type, according to the following definition.

Definition 19 *(i)* Given a faithful representation $G \to Aut(\Lambda)$, where Λ is a free abelian group of even rank $2n$, a G-**Hodge decomposition** is a G-invariant decomposition

$$\Lambda \otimes \mathbb{C} = H^{1,0} \oplus H^{0,1}, \quad H^{0,1} = \overline{H^{1,0}}.$$

(ii)Write $\Lambda \otimes \mathbb{C}$ as the sum of isotypical components

$$\Lambda \otimes \mathbb{C} = \oplus_{\chi \in Irr(G)} U_\chi.$$

Write also $U\chi = W\chi \otimes M\chi$, where $W\chi$ is the irreducible representation corresponding to the character χ, and $M\chi$is a trivial representation whose dimension is denoted $n\chi$.

Write accordingly $V := H1{,}0 = \bigoplus \chi \in \mathrm{Irr}(G) V\chi$, where $V\chi = W\chi \otimes M1{,}0\chi$.

Then the **Hodge type** of the decomposition is the datum of the dimensions

$$\nu(\chi) := dim_{\mathbb{C}} M_{\chi}^{1,0}$$

corresponding to the Hodge summands for non real representations (observe in fact that one must have: $\nu(\chi)+\nu(\bar{\chi})=\dim(M\chi)$).

Remark 20 Given a faithful representation $G \to Aut(\Lambda)$, where Λ is a free abelian group of even rank $2n$, all the G-Hodge decompositions of a fixed Hodge type are parametrized by an open set in a product of Grassmannians. Since, for a non real irreducible representation χ one may simply choose $M1{,}0\chi$ to be a complex subspace of dimension $\nu(\chi)$ of $M\chi$, and for $M\chi = \bar{M}\chi$, one simply chooses a complex subspace $M1{,}0\chi$ of middle dimension. Then the open condition is just that ($\text{(since } M_{\chi}^{0,1} := \overline{M_{\chi}^{1,0}})$) we want

$$M_{\chi} = M_{\chi}^{1,0} \oplus M_{\chi}^{0,1},$$

or, equivalently, $M_{\chi} = M_{\chi}^{1,0} \oplus \overline{M_{\bar{\chi}}^{1,0}}.$.

5.3 The general case where *G* is Abelian

Assume now that G is Abelian, and consider the linear representation

$\rho : G \to \mathrm{GL}(V)$.

We get a splitting

$$V = \oplus_{\chi \in G^{\vee}} V_{\chi} = \oplus_{\chi \in \mathcal{X}} V_{\chi},$$

where $G\vee := Hom(G, \mathbb{C}*) = \{\chi : G \to \mathbb{C}*\}$ is the dual group of characters of G, $V\chi$ is the χ-eigenspace of G (for each $v \in V\chi$, $\rho(g)(v) = \chi(g)v$), and X is the set of characters χ such that $V\chi \neq 0$.

Proposition 21 Assume that an abelian group G acts on a complex torus $X:=V/\Lambda$.

Then the isomorphism class of the group Γ of affine transformations which are lifts of transformations in G determines the real affine type of the action of Γ on $V=(\Lambda\otimes\mathbb{R})$, in particular the above exact sequence determines the action of G up to a real affine isomorphism of $A=(\Lambda\otimes\mathbb{R})/\Lambda$.

Proof By Proposition 17 we may reduce ourselves to the case where G contains no translations.

Step (I): let us first treat the case where G is cyclic, generated by an element g' of order m.

Then we claim that the power gm of a lift g of g' determines the affine type of the transformation g.

In fact, if $g(v)=\alpha v+b$, V splits according to the eigenspaces of the linear map α, which has order exactly m, and we can write $V=V1\oplus V2$, where $V1=\ker(\alpha-Id)$.

Then $g(v1+v2)=(v1+b1)+(\alpha 2(v2)+b2)$, and choosing as the origin in the affine space $V2$ the unique fixed point of $g2:=v2\mapsto\alpha 2(v2)+b2$, we reach the normal form $g(v1+v2)=(v1+b1)+\alpha 2(v2)$. Now, $gm(v1+v2)=(v1+mb1)+v2$, hence gm determines $b1$, and therefore also the normal form of g.

Step (II): split $V=\oplus\chi V\chi$ according the the characters of G, and choose, for each χ, an element $g\chi$ whose image under χ generates $\chi(G)\subset\mathbb{C}*$.

For each χ, as in step I, choose as origin in the affine space $V\chi$ a fixed point for $g\chi$.

Let now g be an arbitrary element in G: then g acts on $V\chi$ by $v\mapsto\chi(g)v+b\chi$.

We only need to show that $b\chi$ is uniquely determined.

This is clear, by step I, if $\chi(g)=1$. Else, there is an integer r such that $\chi(ggr\chi)=1$, hence the affine action of $ggr\chi$on $V\chi$ is uniquely determined, therefore also the one of g.

Remark 22 If G is a general group, we can write $V=\bigoplus \rho V\rho$ where $V\rho$ is irreducible (but the same irreducible representation may occur more than once) and let $H\rho$ be a subgroup such that $\rho(H\rho)=\rho(G)$. Argueing as in step II above, the theorem holds for G if one can prove that the affine action of $H\rho$ on $V\rho$ is uniquely determined up to affine conjugation.

i.e., it suffices to prove the result when V is irreducible and $G\subset Aut(V)$.

Remark 23 In the papers [35, 37] we claimed the validity of Proposition 21 for any finite group G, but the proof was incorrect. Does the result hold for any finite group G, at least in the case of an Abelian variety A? We leave this question open, hoping to return on it.

In the case of generalized hyperelliptic varieties, we require the condition that the action of G is free: this implies the following condition

$$(1)\ \forall g \in G, \exists \chi \in \mathcal{X}, \quad \chi(g) = 1, \Leftrightarrow \cup_{\chi\in\mathcal{X}} \ker(\chi) = G,$$

while we may assume that G contains no translations, i.e., that ρ is injective (equivalently , X spans GV).

With the above in mind, we pass to investigate in more detail the case where G is cyclic, the case where the quotient variety shall be called a Bagnera–de Franchis variety.

5.4 Bagnera–de Franchis varieties of small dimension

In view of the characterization given in Proposition 16, we can see a Bagnera–de Franchis variety as the quotient of one of product type, since the actions of T and g commute (by the property that $\alpha 2\circ\phi=\phi$).

Dealing with appropriate choices of T is the easy part, since, as we saw, the points $t2$ of $T2$ satisfy the property $\alpha 2(t2)=t2$. It suffices to choose $T2\subset A2[*]:=\ker(\alpha 2-\mathrm{Id}A2)$, which is a finite subgroup of $A2$, and then to pick an isomorphism $\psi:T2\rightarrow T1\subset A1$, such that $T1:=Im(\psi)\cap\langle\beta 1\rangle=\{0\}$.

Let us then restrict ourselves to consider Bagnera–de Franchis varieties of product type.

We show now how to further reduce to the investigation of primary Bagnera–de Franchis varieties.

In fact, in the case of a BdF variety of product type, Λ2 is a G -module, hence a module over the group ring

$$R := R(m) := \mathbb{Z}[G] \cong \mathbb{Z}[x]/(x^m - 1).$$

This ring is in general far from being an integral domain, since indeed it can be written as a direct sum of cyclotomic rings, which are the integral domains defined as

$$R_m := \mathbb{Z}[x]/P_m(x).$$

Here $Pm(x)$ is the m-th cyclotomic polynomial

$$P_m(x) = \Pi_{0<j<m,(m,j)=1}(x - \epsilon^j),$$

where ϵ=exp($2\pi i/m$).

Then

$$R(m) = \oplus_{k|m} R_k.$$

The following elementary lemma, together with the splitting of the vector space V as a direct sum of eigenspaces for g, yields a decomposition of $A2$ as a direct product $A2$=⊕$k|mA2,k$ of G-invariant Abelian subvarieties $A2,k$ on which g acts with eigenvalues of order precisely k.

Lemma 24 Assume that M is a module over a ring R=⊕kRk.

Then M splits uniquely as a direct sum M=⊕kMk such that Mk is an Rk-module, and the R-module structure is obtained through the projection R→Rk.

Proof We can write the identity in R as a sum of idempotents 1=Σkek, where ek is the identity of Rk, and $ekej$=0 for$j$≠k.

Then each element w∈M can be written as

$$w = 1w = (\Sigma_k e_k)w = \Sigma_k e_k w =: \Sigma_k w_k.$$

Hence Mk is defined as ekM.

In the situation where we have a primary Bagnera–de Franchis variety Λ2 is a module over the integral domain R:=Rm:=Z[x]/Pm(x),

Since Λ2 is a projective module, a classical result (see [290], lemmas 1.5 and 1.6) is that Λ2 splits as the direct sum Λ2=Rr⊕I of a free module with an ideal I⊂R, and it is indeed free if the class number h(R)=1(to see for which integers m this occurs, see the table in [370], p. 353). To give a complex structure to $A$2:=(Λ2⊗ZR)/Λ2 it suffices to give a decomposition Λ2⊗ZC=V⊕¯V, such that the action of x is holomorphic, which is equivalent to asking that V is a direct sum of eigenspaces Vλ, for λ=ϵj a primitive m-th root of unity.

Writing U:=Λ2⊗ZC=⊕Uλ, the desired decomposition is obtained by choosing, for each eigenvalue λ, a decomposition $U_\lambda = U_\lambda^{1,0} \oplus U_\lambda^{0,1}$ such that $\overline{U_\lambda^{1,0}} = U_{\bar\lambda}^{0,1}$.

The simplest case (see [89] for more details) is the one where I=0,r=1, hence dim(Uλ)=1, so that we have only a finite number of complex structures, depending on the choice of the φ(m)/2 indices j such that Uϵj=$U$1,0ϵj.

The above discussion does however leave open the following question (see [265] for some partial results), which should have an affirmative answer (in the sense that each family of complex structures should contain some Abelian variety).

Question: when is then the complex torus $A$2 an Abelian variety?

Observe that the classification in small dimension is possible thanks to the observation that the Z-rank of R(or of any ideal I⊂R) cannot exceed the real dimension of $A$2: in other words we have

$$\varphi(m) \leq 2(n-1),$$

where φ(m) is the Euler function, which is multiplicative for relatively prime numbers, and satisfies $\varphi(p^r) = (p-1)p^{r-1}$ when p is a prime number.

For instance, when n≤3, then φ(m)≤4, and φ(pr)≤4 iff p=3,5,r=1, or p=2,r≤3.

Observe that the case m=2 was completely described, in view of Proposition 16, hence we may assume that$m$≥3 and ask for the classification for the Bagnera–de Franchis varieties (or manifolds) of order m.

Assuming then $m \geq 3$, if $n \leq 3$, it is only possible to have $m=3,4,6, \varphi(m)=2$, or $m=5,8,10,12, \varphi(m)=4$. The classification is then made easier by the fact that, in the above range for m, R is a P.I.D., hence every torsion free module is free. In particular Λ_2 is a free R-module.

The classification for $n=4$, since we must have $\varphi(m) \leq 6$, is going to include also the case $m=7,9$.

We are not aware of literature dedicated to a precise classification of Bagnera–de Franchis varieties, or generalised hyperelliptic varieties, at least in dimension >3 (see however [264, 357] for results in dimension 3). Observe that the hypothesis that G is a finite group allows to find a G-invariant Hermitian metric on V, hence the affine group Γ extension of Λ by G is a Bieberbach group, and in each dimension we have only a finite number of those.

We end this section mentioning some elementary results which are useful to locate the BdF varieties in the classification theory of algebraic varieties.

Proposition 25 **The Albanese variety of a Bagnera–de Franchis variety $X=A/G$ is the quotient $A_1/(T_1+\langle\beta_1\rangle)$.**

Proof **Observe that the Albanese variety** $H^0(\Omega^1_X)^\vee/Im(H_1(X,\mathbb{Z}))$ of $X = A/G$ **is a quotient of the vector space V_1 by the image of the fundamental group of X (actually of its abelianization, the first homology group $H_1(X,\mathbb{Z})$): since the dual of V_1 is the space of G-invariant forms on A,** $H^0(\Omega^1_A)^G \cong H^0(\Omega^1_X)$.

We also observe that there is a well defined map $X \to A_1/(T_1+\langle\langle\beta_1\rangle\rangle)$, since T_1 is the first projection of T. The image of the fundamental group of X contains the image of Λ, which is precisely the extension of Λ_1 by the image of T, namely T_1. Since we have the exact sequence

$$1 \to \Lambda = \pi_1(A) \to \pi_1(X) \to G \to 1$$

the image of the fundamental group of X is generated by the image of Λ and by the image of the transformation g, which however acts on A_1 by translation by $\beta_1 = [b_1]$.

Remark 26 Unlike the case of complex dimension $n=2$, there are Bagnera–de Franchis varieties $X=A/G$ with trivial canonical divisor, for instance some examples are given by:

(1)any BdF variety which is standard (i.e., $m=2$) and is such that $A2$ has even dimension has trivial canonical divisor, as well as

(2)some BdF varieties $X=A/G=(A1\times A2)/(T\times G)$ whose associated product type BdF variety $Y=(A1\times A2)/G$ is primary and elementary (meaning that $\Lambda2\cong Rm$). For instance, when $m=p$ is an odd prime, there is a primary and elementary BdF variety with trivial canonical divisor if and only if, given the set of integers $\{1, 2, \dots, [\frac{p-1}{2}]\}$, there is a partition of it into two sets whose respective sums are the same: for $m=7$, it suffices to choose the partition $\{1,2\}\cup\{3\}$. This partition yields the sets $\{1,2,4\}$ and $\{3,5,6\}$, which determine $V=H1,0$ as $V=H1,0:=U1\oplus U2\oplus U4$. Whereas, for $m=17$, there are many choices, consisting of the union of two doubletons of the form $\{j,9-j\}$.

We do not further consider the question of determining exactly the case where the divisor KX is trivial, also because this question can be asked also for the more general case where the action of G is not free (see [316] for a classification in dimension $n=3$).

6 ORBIFOLD FUNDAMENTAL GROUPS AND RATIONAL *K(Π,1)*'S

6.1 Orbifold fundamental group of an action

In the previous section we have considered quotients $X=A/G$ of a complex torus $A=V/\Lambda$ by the free action of a finite group G. In this case the affine group Γ fitting into the exact sequence

$1\to\Lambda=\pi1(A)\to\Gamma\to G\to1$

equals $\pi1(X)$.

We have also seen that we have the same exact sequence in the more general case where the action of G is no longer free, and that the group Γ determines in general the affine action of Γ on the real affine space V(resp. on the rational affine space $\Lambda\otimes Q$) and that, once the Hodge type of V is fixed, these varieties are parametrized by a connected complex manifold.

In the case where the action of G is no longer free, we would like to remember the group Γ, in view of its importance, even if it is no longer a 'bona fide' fundamental group. This can be done through a more general correspondence which associates to the pair of X and the group action of G a group Γ which is called the **orbifold fundamental group** and can be defined in many ways (see [96, 135]). Here is one.

Definition 27 Let Z be a 'good' topological space, i.e., arcwise connected and semi-locally 1-connected, so that there exists the universal cover D of Z. Then we have $Z = D/\pi$ where $\pi := \pi_1(Z)$; denote $p: D \to Z$ the quotient projection.

Assume now that a group G acts properly discontinuously on Z, and set $X := Z/G$. Then we define the orbifold fundamental group of the quotient of Z by G as the group of all the possible liftings of the action of G on D, more precisely:

$$\pi_1^{orb}(Z, G) := \{\gamma : \mathcal{D} \to \mathcal{D} | \exists g \in G, \quad s.t.\ p \circ \gamma = g \circ p\}$$

Remark 28 (1) We obtain an exact sequence, called orbifold fundamental group exact sequence,

$$(***)\ 1 \to \pi_1(Z) \to \pi_1^{orb}(Z, G) \to G \to 1$$

since for each $g \in G$ we have a lifting γ of g, because D is the universal cover, so its associated subgroup is the identity, hence the lifting property is ensured. Moreover the lifting is uniquely determined, given base points $z_0 \in Z, y_0 \in D$, with $p(y_0) = z_0$, by the choice of $y \in p^{-1}(\{z_0\})$ with $\gamma(y_0) = y$.

(2) In the case where the action of G is free, then $\pi^{orb}_1(Z,G) = \pi_1(Z/G) = \pi_1(X)$.

(3) If G acts properly discontinuously on Z, then $\Gamma := \pi^{orb}_1(Z,G)$ acts also properly discontinuously on D: in fact the defining property that, for each compact $K \subset D$, $\Gamma(K,K) := \{\gamma | \gamma(K) \cap K \neq \emptyset\}$ is finite, follows since $p(K)$ is compact, and since π acts properly discontinuously.

(4) If the space Z is a $K(\pi,1)$, i.e., a classifying space $Z = B\pi$, then (***) determines the topological type of the G-action. The group Γ acts by

conjugation on π, and this yields an homomorphism $G \to \mathrm{Out}(\pi)=\mathrm{Aut}(\pi)/\mathrm{Inn}(\pi)$, and we just have to recall that the topological action of $\gamma \in \Gamma$ is determined by its action on π, taken up to $\mathrm{Inn}(\pi)$ if we do not keep track of the base point.

In fact there is a bijection between homotopy classes of self maps of Z and homomorphisms of π, taken of course up to inner conjugation (inner conjugation is the effect of changing the base point, and if we do not insist on taking pointed spaces, i.e., pairs $(Z,z0)$ the action of a continuous map on the fundamental group is only determined up to inner conjugation). Clearly a homeomorphism $\varphi: Z \to Z$ yields then an associated element $\pi 1(\varphi) \in \mathrm{Out}(\pi)$.

(5) Even if not useful for computations, one can still interpret the exact sequence (***) as an exact sequence for the fundamental groups of a tower of covering spaces, using the standard construction for equivariant cohomology that we shall discuss later.

Just let $BG=EG/G$ be a classifying space, and consider the free product action of G on $Z \times EG$.

Then we have a tower of covering spaces

$(\mathrm{D} \times EG) \to (Z \times EG) \to (Z \times EG)/G,$

where $\mathrm{D} \times EG$ is simply connected, and is the universal cover of $Y:=(Z \times EG)/G$, such that $\pi_1(Y) \cong \pi_1^{orb}(Z,G)$.

Of course, the above definitions seem apparently unrelated to the fundamental group of X, however we can take the maximal open set $U' \subset \mathrm{D}$ where $\Gamma := \pi orb1(Z,G)$ acts freely. Indeed $U'=p-1(U)$, where U is the open set $U:=Z \setminus \{z \mid Gz:=Stab(z) \neq \{\mathrm{Id}G\}\}$.

The conclusion is that $U'/\Gamma=U/G=:U'' \subset X$, hence we get exact sequences

$$1 \to \pi_1(U) \to \pi_1(U'') \to G \to 1,$$
$$1 \to \pi_1(U') \to \pi_1(U'') \to \Gamma \to 1.$$

We make now the hypothesis that D, hence also Z, X are normal complex spaces, in particular they are locally contractible, and that the actions are holomorphic: then we have surjections

$\pi 1(U'') \to \pi 1(X)$, $\pi 1(U) \to \pi 1(Z)$,

moreover $\pi 1(U') \to \pi 1(\mathrm{D})=\{1\}$, and finally we get

$$\pi_1(Z) = \pi_1(U)/\pi_1(U'),\ \pi_1^{orb}(Z,G) \to \pi_1(X) \to 0.$$

Example 29 Assume that Z, X are smooth complex curves $Z=C$, $X=C'$. Then for each point $p_i \in X \setminus U''$ we get the conjugate $\gamma_i \in \pi_1(U'')$ of a circle around the point p_i, and we have $\ker(\pi_1(U'') \to \pi_1(X)) = \langle\langle\{\gamma_i\}\rangle\rangle$, i.e., the kernel is normally generated by these loops.

Each of these loops γ_i maps to an element $c_i \in G$, and we denote by m_i the order of the element (c_i) in the group G. Then $\gamma_i^{m_i} \in \pi_1(U)$, and it is the conjugate of small circle around a point of $Z \setminus U$, hence it maps to zero in $\pi_1(Z)$. One can see that in this case

$$\pi_1^{orb}(Z,G) = \pi_1(U'')/\langle\langle\{\gamma_i^{m_i}\}\rangle\rangle$$

and the kernel of the natural surjection onto $\pi_1(X)$ is normally generated by the γ_i's.

This means that, if the genus of $X=C'$ is equal to g', then the orbifold fundamental group is isomorphic to the abstract group

$$\pi_{g',m_1,\dots,m_d} := \langle \alpha_1, \beta_1, \dots, \alpha_{g'}, \beta_{g'}, \gamma_1, \dots, \gamma_d | \Pi_1^d \gamma_j \Pi_1^{g'} [\alpha_i, \beta_i] = 1, \gamma_1^{m_1} = \dots = \gamma_d^{m_d} = 1 \rangle.$$

And the orbifold exact sequence is an extension, called Nielsen extension, of type

(NE) $1 \to \pi_g \to \pi_{g',m_1,\dots,m_d} \to G \to 1$.

A similar description holds in general (see [96], definition 4.4 and proposition 4.5, pp. 25–26), at least when Z is a complex manifold.

Remark 30 The notion of orbifold fundamental group has turned out to be quite useful also in real algebraic geometry, where one considers the action of complex conjugation on a real algebraic variety (see [97] for example).

6.2 Rational $K(\pi,1)$'s: basic examples

An important role is also played by **complex Rational** $K(\pi,1)$'s, i.e., quasi projective varieties (or complex spaces) Z such that

$Z=D/\pi$,

where D is a contractible manifold (or complex space) and the action of π on D is properly discontinuous but not necessarily free.

While for a $K(\pi,1)$ we have $H*(\pi,Z)\cong H*(Z,Z)$, $H*(\pi,Z)\cong H*(Z,Z)$, for a rational $K(\pi,1)$, as a consequence of Proposition 46 and of the spectral sequence of Theorem 62, we have $H*(\pi,Q)\cong H*(Z,Q)$ and therefore also $H*(\pi,Q)\cong H*(Z,Q)$.

Typical examples of such rational $K(\pi,1)$'s are:

(1)quotients of a bounded symmetric domain D by a subgroup $\Gamma\subset$Aut(D) which is acting properly discontinuously (equivalently, Γ is discrete); especially noteworthy are the case where Γ is **cocompact**, meaning that X=D/Γ is compact, and the **finite volume** case where the volume of X via the invariant volume form for D is finite.

(2)the moduli space of principally polarized Abelian Varieties, which is an important special case of the above examples;

(3)the moduli space of curves Mg.

For the first example the property is clear (we observed already that such a domain D is contractible).

For the second example it suffices to observe that the moduli space of Abelian Varieties of dimension g, and with a polarization of type ($d1$,$d2$,...,dg) is the quotient of Siegel's upper half space

$$\mathcal{H}_g := \left\{\tau \in \mathrm{Mat}(g, g, \mathbb{C}) | \tau =^t \tau, \mathrm{Im}(\tau) > 0\right\},$$

which is biholomorphic to a bounded symmetric domain of type III in E. Cartan's classification [77], by the properly discontinuous action of the group

$$\mathrm{Sp}(D, \mathbb{Z}) := \left\{M \in Mat(2g, \mathbb{Z}) |^t MDM = D\right\},$$

where

$$D' := \mathrm{diag}(d_1, d_2, \ldots, d_g), D := \begin{pmatrix} 0 & D' \\ -D' & 0 \end{pmatrix}$$

and,

$$M := \begin{pmatrix} \alpha & \beta \\ \gamma & \delta \end{pmatrix}, \quad \tau \mapsto -D'(D'\alpha - \tau\gamma)^{-1}(D'\beta - \tau\delta).$$

In fact, these Abelian varieties are quotients $\mathbb{C}^g/\Lambda$, where Λ is generated by the columns of the matrices $D,-\tau$, and the Hermitian form associated to the real matrix $Im(\tau)^{-1}$ has imaginary part which takes integral values on $\Lambda\times\Lambda$, and its associated antisymmetric matrix in the chosen basis is equal to D.

The quotient $\mathcal{A}_{g,D}' := \mathcal{H}_g/Sp(D,\mathbb{Z})$ is not compact, but of finite volume, and there exist several compactifications which are projective algebraic varieties (see [163, 176, 311]).

6.3 The moduli space of curves

The most useful (and first fully successful) approach to the moduli space of curves of genus g is to view it as a quotient

(**) $\mathfrak{M}_g = \mathcal{T}_g/\mathcal{M}ap_g$

of a connected complex manifold $\mathcal{T}_g$ of dimension $3g-3+a(g)$, called **Teichmüller space**, by the properly discontinuous action of the **Mapping class group** $\mathcal{M}ap_g$ (here $a(0)=3, a(1)=1, a(g)=0, \forall g\geq 2$ is the complex dimension of the group of automorphisms of a curve of genus g). A key result (see [217, 237, 356]) is the

Theorem 31 Teichmüller space $\mathcal{T}_g$ is diffeomorphic to a ball, and the action of $\mathcal{M}ap_g$ is properly discontinuous.

Denoting as usual by π_g the fundamental group of a compact complex curve C of genus g, we have in fact a more concrete description of the mapping class group (see [9, 10, 44, 45, 125, 165, 212, 220] for general results on braid and mapping class groups):

(M) $\mathcal{M}ap_g \cong Out^+(\pi_g)$.

The above superscript + refers to the orientation preserving property.

There is a simple algebraic way to describe the orientation preserving property: any automorphism of a group induces an automorphism of its abelianization, and any inner automorphism acts trivially.

Hence $Out(G)$ acts on G^{ab}, in our particular case $Out(\pi_g)$ acts on $\pi_g^{ab} \cong \mathbb{Z}^{2g}$, and $Out^+(\pi_g) \subset Out(\pi_g)$ is the inverse image of $SL(2g,\mathbb{Z})$.

The above isomorphism (M) is of course related to the fact that C is a $K(\pi_g,1)$, as soon as $g\geq 1$.

As we already discussed, there is a bijection between homotopy classes of self maps of C and endomorphisms of πg, taken up to inner conjugation. Clearly a homeomorphism $\varphi{:}C{\to}C$ yields then an associated element $\pi 1(\varphi)\in \mathrm{Out}(\pi g)$.

Such a homeomorphism acts then on the second homology group $H2(C,\mathrm{Z})\cong \mathrm{Z}[C]$, where the generator $[C]$ corresponds to the orientation associated to the complex structure; the condition $H2(\varphi)$=+1 that φ is orientation-preserving translates into the above algebraic condition on $\psi{:=}\pi 1(\varphi)$. That the induced action ψabon the Abelianization $\pi_g^{ab} \cong \mathbb{Z}^{2g}$, satisfies: $2g(\psi ab)$ acts as the identity on $\wedge^{2g}(\mathbb{Z}^{2g}) \cong \mathbb{Z}.$.

In other words, the image of the product of commutators $\epsilon{:=}\Pi j[\alpha j,\beta j]$ is sent to a conjugate of ϵ.

Turning now to the definition of the Teichmüller space Tg, we observe that it is somehow conceptually easier to give the definition of Teichmüller space for more general manifolds.

6.4 Teichmüller space

Let M be an oriented real differentiable manifold of real dimension $2n$, for simplicity let's assume that M is compact.

Ehresmann [149] defined an **almost complex structure** on M as the structure of a complex vector bundle on the real tangent bundle TMR: namely, the action of $\sqrt{-1}$ on TMR is provided by an endomorphism

$J{:}TM\mathrm{R}{\to}TM\mathrm{R}$,with $J^2{=}{-}Id$.

Equivalently, as done for the complex tori, one gives the decomposition of the complexified tangent bundle TMC:=TMR⊗RC as the direct sum of the i, respectively $-i$ eigenbundles:

$$TM_{\mathbb{C}} = TM^{1,0} \oplus TM^{0,1} \quad \text{where } TM^{0,1} = \overline{TM^{1,0}}.$$

The space AC(M) of almost complex structures, once TMR (hence all associated bundles) is endowed with a Riemannian metric, is a subset of the Fréchet space $H0(M$,C∞(End(TMR))) .

A closed subspace of AC(M) consists of the set C(M) of complex structures: these are the almost complex structures for which there are at each point x local holomorphic coordinates, i.e., C-valued functions z1,...,znwhose differentials span the dual $(TM_y^{1,0})^\vee$ of $TM_y^{1,0}$ for each point *y* in a neighbourhood of *x*.

In general, the splitting

$$TM_{\mathbb{C}}^\vee = (TM^{1,0})^\vee \oplus (TM^{0,1})^\vee$$

yields a decomposition of exterior differentiation of functions as $df = \partial f + \bar{\partial} f$, and a function is said to be holomorphic if its differential is complex linear, i.e., ¯∂*f*=0.

This decomposition *d*=∂+¯∂ extends to higher degree differential forms. The theorem of Newlander and Nirenberg [312], first proven by Eckmann and Frölicher in the real analytic case [144] characterizes the complex structures through an explicit equation:

Theorem 32 (Newlander–Nirenberg) An almost complex structure *J* yields the structure of a complex manifold if and only if it is integrable, which means ¯∂2=0.

Obviously the group of oriented diffeomorphisms of *M* acts on the space of complex structures, hence one can define in few words some basic concepts.

Definition 33

Let D*iff*+(*M*) be the group of orientation preserving diffeomorphisms of *M* , and let C(*M*) be the space of complex structures on *M*. Let D*iff*0(*M*)⊂D*iff*+(*M*) be the connected component of the identity, the so called subgroup of diffeomorphisms which are isotopic to the identity.

Then Dehn [132] defined the mapping class group of *M* as

$$\mathcal{M}ap(M) := \mathcal{D}iff^+(M)/\mathcal{D}iff^0(M),$$

while the Teichmüller space of *M*, respectively the moduli space of complex structures on *M* are defined as

$$\mathcal{T}(M) := \mathcal{C}(M)/\mathcal{D}iff^0(M), \quad \mathfrak{M}(M) := \mathcal{C}(M)/\mathcal{D}iff^+(M).$$

From these definitions follows that

$$\mathfrak{M}(M) = \mathcal{T}(M)/\mathcal{M}ap(M).$$

The simplest examples here are two: complex tori and compact complex curves. In the case of tori a connected component of Teichmüller space (see [100] and also [103]) is an open set Tn of the complex Grassmann Manifold $Gr(n, 2n)$, image of the open set of matrices

$$\{\Omega \in \mathrm{Mat}(2n, n; \mathbb{C}) \mid i^n det(\Omega\overline{\Omega}) > 0\}.$$

This parametrization is very explicit: if we consider a fixed lattice $\Lambda \cong Z^2 n$, to each matrix Ω as above we associate the subspace of $C2n \cong \Lambda \otimes C$ given as

$V = \Omega Cn$,

so that $V \in Gr(n,2n)$ and $\Lambda \otimes C \cong V \oplus {}^{-}V$.

Finally, to Ω we associate the torus

$YV := V/pV(\Lambda) = (\Lambda \otimes C)/(\Lambda \oplus {}^{-}V)$,

$pV: V \oplus {}^{-}V \to V$ being the projection onto the first addendum.

It was observed however by Kodaira and Spencer already in their first article on deformation theory ([248], and volume II of Kodaira's collected works) that the mapping class group $SL(2n,Z)$ does not act properly discontinuously on Tn.

The case of compact complex curves C is instead the one which was originally considered by Teichmüller.

In this case, if the genus g is at least 2, the Teichmüller space Tg is a bounded domain, diffeomorphic to a ball, contained in the vector space of quadratic differentials $H^0(C, \mathcal{O}_C(2K_C)) \cong \mathbb{C}^{3g-3}$ on a fixed such curve C.

In fact, for each other complex structure on the oriented 2-manifold M underlying C we obtain a complex curve C', and there is a unique extremal quasi-conformal map $f: C \to C'$, i.e., a map such that the Beltrami distortion $\mu f := {}^{-}\partial f/\partial f$ has minimal norm (see for instance [217] or [8]).

The fact that the Teichmüller space Tg is diffeomorphic to a ball (see [356] for a simple proof) is responsible for the fact that the moduli space of curves Mg is a rational $K(\pi,1)$.

6.5 Singularities of M*g*, I

Teichmüller's theorem says that, for $g \geq 2$, $Tg \subset C^{3g-3}$ is an open subset diffeomorphic to a ball. Moreover $Mapg \cong Map(C)$ acts properly discontinuously on T*g*, but not freely.

The lack of freeness of this action is responsible of the fact that M*g* is a singular complex space (it is quasi-projective by the results of Mumford and Gieseker, see [302, 304]).

Therefore one has to understand when a class of complex structure *C* on the manifold *M* is fixed by an element $\gamma \in Mapg$. Since γ is represented by the class of an orientation preserving diffeomorphism $\varphi: M \to M$, it is not difficult to see that this situation means that the diffeomorphism φ carries the given complex structure to itself, or, in other words, the differential of φ preserves the splitting of the complexified tangent bundle of $M, TM_{\mathbb{C}} = TM^{1,0} \oplus T\bar{M}^{0,1}$.

This is then equivalent to $\bar{\partial}\varphi = 0$, i.e., φ is a biholomorphic map, i.e., $\varphi \in \mathrm{Aut}(C)$.

Conversely, let $\gamma \in \mathrm{Aut}(C)$ be a nontrivial automorphism. Consider then the graph of γ, $\Gamma\gamma \subset C \times C$, and intersect it with the diagonal $\Delta \subset C \times C$: their intersection number must be a nonnegative integer, because intersection points of complex subvarieties carry always a positive local intersection multiplicity.

This argument was used by Lefschetz to prove the following

Lemma 34 (Lefschetz' lemma) If we have an automorphism $\gamma \in \mathrm{Aut}(C)$ of a projective curve of genus $g \geq 2$, then γ is homotopic to the identity iff it is the identity:

$$\gamma \underset{h}{\sim} Id_C \Leftrightarrow \gamma = Id_C$$

Proof The self intersection of the diagonal Δ2 equals the degree of the tangent bundle to *C*, which is $2-2g$. If γ is homotopic to the identity, then $\Gamma\gamma \cdot \Delta = 2-2g < 0$, and this is only possible if $\Gamma\gamma = \Delta$, i.e., iff $\gamma = idC$.

In particular, since *C* is a classifying space $B\pi g$ for the group πg, $\gamma: C \to C$ is such that the homotopy class of γ is determined by the conjugacy class of $\pi 1(\gamma): \pi 1(C, y0) \to \pi 1(C, \gamma \cdot y0)$. Hence we get an injective group homomorphism

$$\rho_C: \operatorname{Aut}(C) \to \frac{\operatorname{Aut}(\pi_g)}{\operatorname{Inn}(\pi_g)} = \operatorname{Out}(\pi_g).$$

Actually, since a holomorphic map is orientation preserving, we have that ρC: $\operatorname{Aut}(C) \to \operatorname{Out}^+(\pi_g) = \operatorname{Map}_g \cong \frac{\mathcal{D}iff^+(C)}{\mathcal{D}iff^0(C)}$.

The conclusion is then that curves with a nontrivial group of automorphisms Aut(C)≠{IdC} correspond to points of Tg with a non trivial stabilizer for the action of Mapg.

Remark 35 There is of course the possibility that the action of Mapg is not effective, i.e., there is a normal subgroup acting trivially on Tg. This happens exactly in genus g=2, where every curve is hyperelliptic and the normal subgroup of order 2 generated by the hyperelliptic involution is exactly the kernel of the action.

The famous theorem of Hurwitz gives a precise upper bound |Aut(C)|≤84(g−1) for the cardinality of the stabilizer Aut(C) of C (that this is finite follows also by the proper discontinuity of the action).

Now, coming to Sing(Mg), we get that Mg is locally analytically isomorphic to a quotient 'singularity' C^{3g-3}/Aut(C) (indeed, as we shall see, to the quotient singularity of $H$1(ΘC)/Aut(C) at the origin).

Now, given a quotient singularity Cn/G, where G is a finite group, then by a well known lemma by H. Cartan [78] we may assume that G acts linearly, and then a famous theorem by Chevalley [124] and Shephard and Todd [338] says that the quotient Cn/G is smooth if and only if the action of G⊂GL(n,C) is generated by **pseudo-reflections**, i.e., matrices which are diagonalizable with (n−1) eigenvalues equal to 1, and the last one a root of unity.

Theorem 36 Let G⊂GL(n,C), let Gpr be the normal subgroup of G generated by pseudo reflections, and let Gq be the factor group. Then

(1)the quotient Cn/Gpr is smooth,

(2)We have a factorization of the quotient map Cn→Cn/G as Cn→(Cn/Gpr)≅Cn→Cn/Gq and Cn/G is singular if G≠Gpr.

This result is particularly interesting in this situation, since the only pseudo-reflection occurs for the case of the hyperelliptic involution in genus g=3.

Teichmüller theory can be further applied in order to analyse the fixed loci of finite subgroups G of the mapping class group (see [96, 237, 356])

Theorem 37 (Refined Nielsen realization) Let $G \subset \mathrm{Map}_g$ be a finite subgroup. Then $Fix(G) \subset \mathcal{T}_g$ is a non empty complex manifold, diffeomorphic to a ball.

Remark 38 (1) The name Nielsen realization comes from the fact that Nielsen [313–315] conjectured that any topological action of a finite group G on a Riemann surface could be realized as a group of biholomorphic automorphisms if and only if the action would yield an embedding $\rho: G \to \mathrm{Map}_g$.

(2) The question can be reduced to a question of algebra, by the following observation.

The group π_g, for $g \geq 2$, has trivial centre: in fact, two commuting hyperbolic transformations in $PSL(2,\mathbb{R})$ can be simultaneously written as Möbius transformations of the form $z \mapsto \lambda z, z \mapsto \mu z$, hence if the first transformation were in the centre of $\pi_g \subset PSL(2,\mathbb{R})$, the group π_g would be abelian, a contradiction.

Then $\pi_g \cong Inn(\pi_g)$ and from the exact sequence

$1 \to \pi_g \cong Inn(\pi_g) \to \mathrm{Aut}(\pi_g) \to \mathrm{Out}(\pi_g) \to 1$

and an injective homomorphism $\rho: G \to \mathrm{Map}_g = \mathrm{Out}^+(\pi_g)$ one can pull-back an extension[5]

$1 \to \pi_g \to \hat{G} \to G \to 1.$

(4) The main question is to show that the group $\hat{G}$ is isomorphic to an orbifold fundamental group of the form $\pi_{g',m_1,\ldots,m_d}$. Because then we have a Nielsen extension, and the epimorphism $\mu: \pi_{g',m_1,\ldots,m_d} \to G \to 1$ combined with Riemann's existence theorem finds for us a curve C over which G acts with the given topological type: such a curve C is given by the ramified covering of a curve C' of genus g', branched on d points $y_1,\ldots,y_d$ and with monodromy μ.

(5) Since π_g is torsion free (hyperbolic elements in $PSL(2,\mathbb{R})$ have infinite order), we see that the branch points correspond to conjugacy classes of cyclic subgroups of finite order in $\hat{G}$, and that their order m_i in $\hat{G}$ equals the order of their image in G.

In turn, the question can be reduced to showing that there is a differentiable action of the group G on the curve C inducing the topological action ρ. There is the following

Lemma 39 Given $\rho:G\to Mapg=\mathrm{Diff}+(C)/\mathrm{Diff}^0(C)$, if there is a homomorphism $\psi:G\to \mathrm{Diff}+(C)$ whose projection to $Mapg$ is ρ, then there is a complex curve with a G action of topological type ρ.

Proof In fact, by Cartan's lemma [78], at each fixed point $x\in C$, there are local coordinates such that in these coordinates the action of the stabilizer Gx of x is linear (in particular, if we assume that the action of G is orientation preserving, then $Gx\subset SO(2,\mathrm{R})$, and Gx is a cyclic group of rotations).

Therefore, for each $\gamma\in G$, $\gamma\neq 1G$, the set of fixed points $Fix(\gamma)$, which is closed, is either discrete (hence finite), or it consists of a discrete set plus a set (the points $x\in Fix(\gamma)$ where the derivative is the identity) which is open and closed. Since C is connected, the only possible alternative is that, for $\gamma\neq 1G$, the set of fixed points $Fix(\gamma)$ is finite.

Now, it is easy to see that the quotient $C/G=:C'$ is a smooth Riemann surface and we can take a complex structure on C' and lift it to C, so that $C'=C/G$ as complex curves.

Remark 40 In general Marković has proven that the mapping class group cannot be realized, as hoped by Thurston, as a subgroup of the group of diffeomorphisms of a Riemann surface of genus g. There are other recent results concerning non finite groups (see [41]).

Remark 41 (1) It is interesting to observe that one can view the orbifold fundamental group, in the case of curves, as the quotient of a bona fide fundamental group.

Let TC be the tangent bundle of the complex curve, and assume that G has a differentiable action on C, so that G acts on TC, preserving the zero section. We can further assume, by averaging a Riemannian metric onTC, that G preserves the sphere bundle of TC,

$SC:=\{v\in TC||v|=1\}$.

Since each stabilizer Gx is a cyclic group of rotations, the conclusion is that G acts on SC by a free action, and we get an exact sequence

(*Seifert*) $1\to\pi 1(SC)\to\pi 1(SC/G)\to G\to 1$.

It is well known (see e.g. [105]) that the fundamental group of the fibre bundle SC is a central extension

$1 \to Z \to \pi 1(SC) \to \pi g \to 1,$

and is the quotient of the direct product Z×F2g by the subgroup normally generated by the relation

$$c^{2g-2}\Pi_1^g[\alpha_i,\beta_i]=1,$$

where c is a generator of the central subgroup $\langle c\rangle \cong Z$.

Taking the quotient by the subgroup $\langle c\rangle$ generated by c one obtains the orbifold group exact sequence. The fibration $SC/G \to C' = C/G$ is a so-called Seifert fibration, with fibres all homemorphic to $S1$, but the fibres over the branch points yj are multiple of multiplicity mj.

Now, an injection $G \to$Out+(πg) determines an action of G on $\pi 1(SC)$ (sending c to itself), an extension of the type we denoted (Seifert), and a topological action on the classifying space SC.

(2) The main point of the proof of the Nielsen realization is however based on analysis, and, more precisely, on the construction of a Morse function f on Tg of class $C1$ which is G-invariant, strictly convex and proper, and therefore has only one minimum.

The unique minimum must be G-invariant, hence we get a point C in $Fix(G)$. Since the function is strictly convex and proper, it does not have other critical points, and it has only one minimum: therefore Morse theory tells that Tg is diffeomorphic to a ball.

Now, in turn, the locus $Fix(G)$ is a non empty submanifold, and the group G acts nontrivially on the normal bundle of $Fix(G)$ in Tg.

The same analysis applies to the restriction of the Morse function to any connected component Y of $Fix(G)$. The function has exactly one critical point D, which is an absolute minimum on Y, hence Y is also diffeomorphic to a ball.

To show that $Fix(G)$ is connected it suffices to show that $C=D$, i.e., that D is also a critical point on the whole space Tg. But the derivative of f at D is a G-invariant linear functional, vanishing on the subspace of G-invariants: therefore the derivative of f at D is zero.

Hence $Fix(G)=Y$, as well as Tg, is diffeomorphic to a ball.

(3) It may happen that $Fix(G)$ may be also the fixed locus for a bigger subgroup $H \subset Mapg$. We shall see examples later in Examples 189, 190. To determine when this happens is an interesting question, fully answered by Cornalba in the case where G is a cyclic group of prime order [127] (this result is the key to understanding the structure of the singular locus Sing(Mg)).

A partial answer to the question whether G is not a **full** subgroup (i.e., $\exists H \supset G$, $Fix(H)=Fix(G)$ was given by several authors, who clarified the orbifold fundamental groups associated to the quotients $C \to C/G \to C/H$ [275, 327, 343].

6.6 Group cohomology and equivariant cohomology

The roots of group cohomology go back to Jacobi and to the study of periodic meromorphic functions as quotients of quasi-periodic holomorphic functions; these functions, which can more generally be taken as vector-valued functions s:Cn→Cm, are solutions to the functional equation

$s(x+\gamma)=f\gamma(x)s(x)$,

where Λ⊂Cn is a discrete subgroup, ∀γ∈Λ⊂Cn,x∈Cn,$f\gamma$:Cn→GL(m,C), and clearly $f\gamma$ satisfies then the cocycle condition

$$f_{\gamma+\gamma'}(x) = f_\gamma(x+\gamma')f_{\gamma'}(x), \quad \forall\gamma, \gamma' \in \Lambda, \quad \forall x \in \mathbb{C}^n.$$

Later on, the ideas of Jacobi were generalised through the concept of vector bundles, since the cocycle $f\gamma(x)$can be used to constructing a vector bundle on the quotient manifold Cn/Λ taking a quotient of the trivial bundle Cn×Cm→Cn by the relation

$$(x, v) \sim (x+\gamma, f_\gamma(x)v),$$

and the cocycle relation just says that the above is an equivalence relation. The main difficulty classically consisted in simplifying as much as possible the form of these cocycles, replacing $f\gamma(x)$ by an equivalent one $\phi(x+\gamma)f\gamma(x)\phi(x)-1$, for ϕ:Cn→GL(m,C).

The construction can be vastly generalized by viewing any bundle on a classifying space $Y=B\Gamma$, quotient of a contractible space $X=E\Gamma$, as the quotient of the trivial bundle $X \times F$ by an action which covers the given action of G on X, hence such that there exists $f\gamma : X \to Homeo(F)$ with

$\gamma(x,v)=(\gamma(x), f\gamma(x)(v)) \Rightarrow \forall \gamma,\gamma', \forall x \in X, f\gamma'\gamma(x)=f\gamma'(\gamma(x))f\gamma(x)$.

As remarked by David Mumford [309], classical mathematics led to concrete understanding of 1-cocycles and 2-cocycles, hence of first and second cohomology groups. But the general machinery of higher cohomology groups is harder to understand concretely.

As always in mathematics, good general definitions help to understand what one is doing, but explicit calculations remain often a hard task. Let us concentrate first on the special case where we look at the singular cohomology groups of a classifying space.

In the case of a classifying space $B\Gamma$, we know that for each ring of coefficients R,

$H*(\Gamma,R)=H*(B\Gamma,R)$,

and for this it suffices indeed that

$H*(\Gamma,\mathbb{Z})=H*(B\Gamma,\mathbb{Z})$.

Which side of the equation is the easier one to get hold of? The answer depends of course on our knowledge of the group Γ and of the classifying space $B\Gamma$, which is determined only up to homotopy equivalence.

For instance, we saw that an incarnation of $B(\mathbb{Z}/2)$ is given by the inductive limit $\mathbb{P}^{\infty}_{\mathbb{R}} := \lim_{n\to\infty} \mathbb{P}^{n}_{\mathbb{R}}$.

A topologist would just observe that P∞R is a CW-complex obtained adding a cell σn in each dimension n, and with boundary map

$$\partial(\sigma_{2n}) = 2\sigma_{2n-1}, \quad \partial(\sigma_{2n-1}) = 0 \cdot \sigma_{2n-2}, \quad \forall n \geq 1.$$

Therefore $H2i(\mathbb{Z}/2,\mathbb{Z})=0, H2i-1(\mathbb{Z}/2,\mathbb{Z})=\mathbb{Z}/2, \forall i \geq 1$, and taking the dual complex of cellular cochains one gets:

$$H^{2i}(\mathbb{Z}/2,\mathbb{Z}) = \mathbb{Z}/2,\ H^{2i-1}(\mathbb{Z}/2,\mathbb{Z}) = 0,\ \forall i \geq 1.$$

There are two important ways how this special calculation generalizes:

(i)a completely general construction of a CW complex which is a classifying space $B\Gamma$ for a finitely presented group Γ,

(ii)a completely algebraic definition of group cohomology.

(i)In fact, assume we are given a finite presentation of Γ as

$\Gamma=\langle x1,...,xn|R1(x),...Rm(x)\rangle$,

which means that Γ is isomorphic to the quotient of the free group Fn with generators $x1,...,xn$ by the minimal normal subgroup R containing the words $Rj\in$Fn (R is called the subgroup of relations, and it is said to be normally generated by the relations Rj).

Then the standard construction of the 2-skeleton of $B\Gamma$ is the CW-complex $B\Gamma2$, of dimension two, which is obtained by attaching, to a bouquet of n circles (which correspond to the generators $x1,...,xn$), m 2-cells whose respective boundary is the closed path corresponding to the word $Rj,\forall j=1,...,m$.

For instance,

$$\mathbb{Z}/2 = \mathbb{Z}/(2\mathbb{Z}) = \langle\langle x_1 | x_1^2\rangle\rangle,$$

and the above procedure produces $\mathbb{P}^2_{\mathbb{R}}$ as 2-skeleton.

However, the universal cover of $\mathbb{P}^2_{\mathbb{R}}$ is the sphere $S2$, which has a homotopy group $\pi 2(S^2)\cong$Z, and we know that the higher ($i\geq 2$) homotopy groups $\pi i(X)$ of a space equal the ones of its universal cover. Then a 3-cell is attached to $\mathbb{P}^2_{\mathbb{R}}$, obtaining $\mathbb{P}^3_{\mathbb{R}}$, and one continues to attach cells in order to kill all the homotopy groups. The same is done more generally to obtain $B\Gamma$ from the CW-complex $B\Gamma2$: 3-cells are attached in order to kill the second homotopy groups, and one then obtains $B\Gamma3$; one obtains then $B\Gamma4$ adding 4-cells in order to kill $\pi 3(B\Gamma3)$, and so on.

The disadvantage of this construction is that from the third step onwards it is no longer so explicit, since calculating homotopy groups is difficult.

An easy but important remark used in the proof of a theorem of Gromov (see [91, 196]) is that when the group π has **few relations**, which more precisely means here $n\geq m+2$, then necessarily there is an integer $b$with $0\leq b\leq m$ such that the rank of $H1(B\pi,$Z$)$ equals $n-m+b$, while rank($H2(B\pi,$Z$)$)$\leq b$: since by the above construction $H2(B\pi 2,$Z$)$ is the kernel of a linear map $\partial:Zm \to H1(B\pi 1,Z)=Zn$, whose rank is denoted by $m-b$, and we have a surjection $H2(B\pi 2,Z) \to H2(B\pi,Z)$.

Hence, if we consider the cup product bilinear map

$$\cup : H^1(B\pi, \mathbb{Z}) \times H^1(B\pi, \mathbb{Z}) \to H^2(B\pi, \mathbb{Z}),$$

for each $\psi \in H1(B\pi,Z)$ the linear map $\phi \mapsto \psi \cup \phi$ has a kernel of rank $\geq n-m \geq 2$. This means that each element $\psi \in H1(B\pi,Z)$ is contained in an isotropic subspace (for the cup product map) of rank at least 2.

It follows that if the fundamental group of a space X, $\Gamma:=\pi 1(X)$ admits a surjection onto π, then the induced classifying continuous map $\phi:X \to B\pi$ (see 11) has the property

- that its induced homomorphism on first cohomology

$$H^1(\phi) : H^1(B\pi, \mathbb{Z}) \to H^1(X, \mathbb{Z})$$

is injective (since $H1(\phi):H1(X,Z) \to H1(B\pi,Z)$ is surjective)

- the image W of $H1(\phi)$ is such that each element $w \in W$ is contained in an isotropic subspace of rank ≥ 2.

We shall explain Gromov's theorem, for the case where X is a Kähler manifold, in the next section.

But let us return to the algebraic point of view (see for instance [223], pp. 355–362, also [63]).

(ii): Assume now that G is a group, so that we have the group algebra A:=Z[G]. Recall the functorial definition of group cohomology and of group homology, which gives a high brow explanation for a rather concrete definition we shall give later.

Definition 42 Consider the category of modules M over the group ring A:=Z[G] (i.e., of abelian groups over which there is an action of G, also called G-modules).

Then there are two functors, the functor of **invariants**

$$M \mapsto M^G := \{w \in M | g(w) = w \quad \forall g \in G\} = \cap_{g \in G} \ker(g - Id),$$

associating to each module M the submodule of elements which are left fixed by the action of G, and the functor of **co-invariants**

$$M \mapsto M_G := M/(\Sigma_{g\in G} Im(g - Id)),$$

associating to each module *M* the minimal quotient module on which the action of *G* becomes trivial.

Both *MG* and *MG* are trivial *G*-modules, and one defines the cohomology groups *Hi*(*G*,*M*) as the derived functors of the functor of invariants, while the homology groups *Hi*(*G*,*M*) are the derived functors of the functor of co-invariants.

The relation of these concepts to homological algebra is furnished by the following elementary lemma.

Lemma 43 Let Z be the trivial *G*-module, i.e., we consider the abelian group Z with module structure such that every *g*∈*G* acts as the identity.

Then we have canonical isomorphisms

$$Hom_{\mathbb{Z}[G]}(\mathbb{Z}, M) \cong M^G, \quad M \otimes_{\mathbb{Z}[G]} \mathbb{Z} \cong M_G,$$

given by the homomorphisms *f*↦*f*(1), respectively *w*⊗*n*↦[*nw*].

In particular, we have

$$H^i(G, M) = Ext^i_{\mathbb{Z}[G]}(\mathbb{Z}, M), \quad H_i(G, M) = Tor_i^{\mathbb{Z}[G]}(\mathbb{Z}, M).$$

We illustrate now the meaning of the above abstract definition for the case of group cohomology.

One can construct an explicit free resolution (a projective resolution would indeed suffice) for the trivial A-module *M*=Z

Ln→*Ln*−1→⋯→*L*1→*L*0→Z→0.

Then for each A-module *M* the cohomology groups *Hn*(*G*,*M*) are computed as the cohomology groups of the complex

Hom(*L*0,*M*)→Hom(*L*1,*M*)→⋯→Hom(*Ln*−1,*M*)→Hom(*Ln*,*M*)→⋯.

While the homology groups are calculated as the homology groups of the complex

→*M*⊗Z[*G*]*Li*+1→*M*⊗Z[*G*]*Li*→*M*⊗Z[*G*]*Li*−1→⋯.

For instance, if G is a cyclic group of order m, then $A=Z[x]/(xm-1)$ and the free resolution of Z is given by free modules of rank one: the homomorphisms are the augmentation $\epsilon:L0\to Z$ (defined by setting $\epsilon(\Sigma_g a_g \cdot g):=\Sigma gag$) and the scalar multiplications:

$(x-1):L2n+1\to L2n$, $(1+x\cdots+x^{m-1}):L2n+2\to L2n+1$,

in view of the fact that $(x-1)(1+x\cdots+x^{m-1})=xm-1\equiv 0\in A$.

The case $m=2$ we saw before is a very special case, since then $A=Z\oplus Zx$, with $x2=1$, hence on the trivial module $Z=\mathrm{Hom}Z[G](Z[G],Z)$ $(1-x)$ acts as multiplication by 0, $1+x$ acts as multiplication by 2. But the same pattern happens for all m: $(1-x)$ acts as multiplication by 0, $1+x\cdots+xm-1$ acts as multiplication by m.

In particular, the homology groups can be calculated as

$H2i(Z/m,Z)=0,i\geq 1,H2i+1(Z/m,Z)\cong Z/m,i\geq 0.$

For more general groups there is a general complex, called the bar-complex, which yields a resolution of the A-module Z.

It is easier to first give the concrete definition of the cohomology groups (see also [180]).

Definition 44 Given a group G and a $A:=Z[G]$-module M, one defines the group of i-cochains with values in M as:

$$C^i(G,M) := \{f : G^{i+1} \to M | f(\gamma g_0,\dots,\gamma g_i) = \gamma f(g_0,\dots,g_i) \quad \forall \gamma \in G\},$$

and the differential $di:Ci(G,M)\to Ci+1(G,M)$ through the familiar formula

$$df(g_0,\dots,g_{i+1}) := \Sigma_0^{i+1}(-1)^j f(g_0,\dots,\hat{g}_j,\dots,g_{i+1}).$$

Then the groups $Hi(G,M)$ are defined as the cohomology groups of the complex of cochains, that is,

$Hi(G,M)=\ker(di)/Im(di-1).$

Remark 45 (1) In group theory one gives a different, but equivalent formula, obtained by considering a cochain as a function of i instead of $i+1$ variables, as follows:

$\varphi(g1,\dots,gi):=f(1,g1,g1g2,\dots,g1g2\dots gi).$

One observes in fact that, in view of equivariance of *f* with respect to left translation on *G*, giving *f* is equivalent to giving φ.

Then the formula for the differential becomes asymmetrical

$$d\varphi(g_1,\ldots,g_{i+1}) = g_1\varphi(g_2,\ldots,g_{i+1}) - \varphi(g_1g_2,g_3,\ldots,g_{i+1}) + \varphi(g_1,g_2g_3,\ldots,g_{i+1})\ldots(-1)^i\varphi(g_1,g_2,g_3,\ldots,g_i).$$

(2) The second formula is more reminiscent of the classical formulae, and indeed, for i=0, yields

$$H^0(G,M) = \{x \in M | gx - x = 0 \quad \forall g \in G\} = M^G,$$
$$Z^1(G,M) = \{\varphi = \varphi(g) | g_1\varphi(g_2) - \varphi(g_1g_2) + \varphi(g_1) = 0 \quad \forall g_1, g_2 \in G\},$$
$$B^1(G,M) = \{\varphi = \varphi(g) | \exists x \in M, \varphi(g) = gx - x\},\ H^1(G,M) = Z^1(G,M)/B^1(G,M).$$

(3) Hence, if M is a trivial G-module, then $B1(G,M)=0$ and $H1(G,M)=\mathrm{Hom}(G,M)$.

We have been mainly interested in the case of Z-coefficients, however if we look at coefficients in a field k, we have:

Proposition 46 Let G be a finite group, and let k be a field such that the characteristic of k does not divide $|G|$. Then

$H^i(G,k)=0\forall i\geq 1$.

***Proof* If M is a $k[G]$-module, we observe first that**

$$M^G \cong Hom_{\mathbb{Z}[G]}(\mathbb{Z},M) \cong Hom_{k[G]}(k,M),$$

and then that the functor $M\mapsto MG$ is exact, because MG is a direct summand of M (averaging $v \mapsto \frac{1}{|G|}\sum_{g\in G} gv$ yields a projector with image MG). Hence $Hi(G,M)=0\ \forall i\geq 1$.

6.7 Group homology, Hopf's theorem, Schur multipliers

To give some explicit formula in order to calculate homology groups, we need to describe the bar-complex, which gives a resolution of the trivial module Z.

In order to relate it to the previous formulae for group cohomology, we should preliminary observe that the group algebra Z[*G*] can be thought of as the subalgebra of the space of functions $a{:}G\to Z$ generated by the characteristic functions of elements of *G* (so that $g\in Z[G]$ is the function such that $g(x)=1$ for $x=g$, else $g(x)=0, x\neq g$. As well known, multiplication on the group algebra corresponds to convolution of the corresponding functions, $f1*f2(x)=\int Gf1(xy-1)f2(y)dy$.

It is then clear that the tensor product $Z[G]\otimes ZM$ yields a space of *M*-valued functions on *G* (all of them, if the group *G* is finite), and we get a space of Z-valued functions on the *i*-th Cartesian product $Gi+1$ by considering the (i+1)-fold tensor product

$Ci{:}=Z[G]\otimes ZZ[G]\otimes\cdots\otimes ZZ[G]$,

with Z[*G*]-module structure given by

$g(x0\otimes\cdots\otimes xi){:}=g(x0)\otimes\cdots\otimes xi$.

Definition 47 The bar-complex of a group *G* is the homology complex given by the free Z[*G*]-modules *Ci* (a basis is given by

$\{(g1,\ldots,gi){:}=1\otimes g1\otimes\cdots\otimes gi\}$)

and with differential $di{:}Ci\to Ci-1$ (obtained by dualizing the previously considered differential for functions), defined by

$$d_i(g_1,g_2,\ldots,g_i) := g_1(g_2,\ldots,g_i) + \Sigma_1^{i-1}(-1)^j(g_1,\ldots g_{j-1},g_jg_{j+1},\ldots,g_i) \\ +(-1)^i(g_1,g_2,\ldots,g_{i-1}).$$

The augmentation map $\epsilon{:}C0=Z[G]\to Z$ shows then the the bar-complex is a free resolution of the trivial Z[*G*]-module Z.

As a consequence, the homology groups $Hn(G,Z)$ are then the homology groups of the complex $(Cn\otimes Z[G]Z,dn)$, and one notices that, since $Z[G]\otimes Z[G]Z\cong Z$, (observe in fact that $g\otimes 1=1\otimes g1=1\otimes 1$), then$Cn\otimes Z[G]Z$ is a trivial Z[*G*]-module and a free Z-module with basis $\{(g1,\ldots,gn)\}$.

The algebraic definition yields the expected result for i=1.

Corollary 48 $H1(G,\mathrm{Z})=G^{ab}=G/[G,G]$, $[G, G]$ being as usual the subgroup generated by commutators.

Proof **First argument, from algebra: we have**

$$\oplus_{g_1, g_2 \in G}\mathbb{Z}(g_1, g_2) \to \oplus_{g \in G}\mathbb{Z}g \to \mathbb{Z}$$

but the second homomorphism is zero, hence

$$H_1(G, \mathbb{Z}) = (\oplus_{g \in G}\mathbb{Z}g)/\langle g_2 - g_1 g_2 + g_1 \rangle = G^{ab},$$

since we have then the equivalence relation $g1g2 \equiv g1+g2$.

Second argument, from topology: $H1(X,\mathrm{Z})$ is the abelianization of the fundamental group, so we just apply this to the case of $X=BG$, since $\pi 1(BG) \cong G$.

The machinery of algebraic topology is also useful in order to show the following 'duality' statement.

Theorem 49

If G is a finite group there is an isomorphism between the group $H2(G,\mathrm{Z})$ and the group of 'Schur multipliers' $H2(G,\mathrm{C}*)$.

Proof **Let us consider the exact sequence of groups**

$0 \to \mathrm{Z} \to \mathrm{C} \to \mathrm{C}* \to 1$,

and use that $Hi(G,\mathrm{C})=0$ for $i \geq 1$ (see [223]).

Hence we have an isomorphism $H2(G,\mathrm{C}*) \cong H3(G,\mathrm{Z})$.

Now, by the universal coefficient formula, the torsion subgroup of $Hn(X,\mathrm{Z})$ equals the torsion subgroup of $Hn+1(X,\mathrm{Z})$. Apply this observation to the case $X=BG$, $n=2$, remarking that the groups $Hi(G,\mathrm{Z})$, hence also the groups $Hi(G,\mathrm{Z})$, are torsion abelian groups for $i \geq 1$. Indeed, $Hi(G,\mathrm{R})=Hi(BG,\mathrm{R})=Hi(EG,\mathrm{R})G=0$). ▯

The calculation of the above group $H2(G,\mathrm{Z})$ was achieved by Hopf [215], and we shall sketch the underlying topological idea.

Theorem 50 **(Hopf) Assume that we have a presentation of a group *G* as the quotient *G*=*F*/*R* of a free group. Then**

$H2(G,\mathrm{Z})=([F,F] \cap R)/[F,R]$.

***Proof* (*Sketch*)**

Proof I (currently fashionable argument)

To the exact sequence of groups

$1 \to R \to F \to G \to 1$

there correspond continuous maps of classifying spaces. These maps are only well defined up to homotopy, and using a small trick we can arrange the corresponding maps to be the standard inclusion of a fibre into a fibre bundle.

More precisely, let $BG=EG/G$ be a classifying space for G, and $BF=EF/F$ be a classifying space for F: since the group is free, BF shall be a bouquet of circles. Moreover, since R is a subgroup of F, its classifying space BR is a covering of BF, indeed it is $BR=EF/R$, and is a CW-complex of dimension 1 (whence, the theorem of Nielsen that a subgroup R of a free group F is also free).

The surjection $F \to G$ yields an action of F on EG, which restricts to a trivial action of R.

We use the construction of equivariant cohomology to obtain a new classifying space

$B'F := (EF \times EG)/F.$

The projection of the product $(EF \times EG) \to EG$ yields a fibre bundle $(EF \times EG)/F \to EG/G$ with fibre EF / R.

Hence an associated sequence of continuous maps

$BR=EF/R \to (EF \times EG)/F=B'F \to EG/G=BG,$

associated to the fibre bundle $B'F \to BG$ and with 1-dimensional fibre BR.

We obtain associated maps of first homology groups

$H1(BR,\mathbb{Z})=R/[R,R] \to H1(BF,\mathbb{Z})=F/[F,F] \to H1(BG,\mathbb{Z})=G/[G,G],$

and since $BR,B'F$ have vanishing i-th homology groups for $i \geq 2$, application of the Serre spectral sequence for fibre bundles [336] shows that we have the exact sequence

$0 \to H2(G,\mathbb{Z}) \to H1(BR,\mathbb{Z})F=(R/[R,R])F \to F/[F,F] \to G/[G,G] \to 0.$

In the above exact sequence appears the group of co-invariants $(R/[R,R])F$, which is the quotient of R / $[R, R]$ by the relations of the form $frf-1=r$, $\forall f\in F$, $r\in R$, or, equivalently, by the subgroup $[F, R]$ generated by the commutators $frf-1r-1$. In other words,

$(R/[R,R])F=R/[F,R]$.

Now, from the exact sequences

(∗) $0\to H2(G,\mathrm{Z})\to R/[F,R]\to \mathrm{K}\to 0$

and

$0\to \mathrm{K}\to F/[F,F]\to G/[G,G]\to 0$,

since $\ker(F\to G/[G,G])=R[F,F]$, we infer that

$\mathrm{K}=(R[F,F])/[F,F]=R/(R\cap[F,F])$,

where the last equality follows from the third isomorphism theorem for groups.

Since $[F,R]\subset(R\cap[F,F])$, Hopf's theorem follows from (*) above.

Proof II (direct argument): follows from the following lemma, since the kernel of $R/[F,R]\to F/[F,F]$ clearly equals $(R\cap[F,F])/[F,R]$. ▯

The relation between algebra and geometry in Hopf's theorem is further explained in lemma 2.5 of [116].

Lemma 51 **Let *BG* be a CW-complex which is a classifying space for the group *G=F/R*, such that**

(i)its 1-skeleton $BG1$ has all the 1-cells in bijective correspondence with a set of generators of F,

(ii)its 2-skeleton $BG2$ has all the 2-cells in bijective correspondence with a set of generators of R (hence$R$ / $[R, R]$ is isomorphic to the relative homology group $H2(BG2,BG1,\mathrm{Z})$).

Then the following exact sequence of relative homology

$0\to H2(BG,\mathrm{Z})\to H2(BG,BG1,\mathrm{Z})\to H1(BG1,\mathrm{Z})\to H1(BG,\mathrm{Z})\to 0$

is isomorphic to

$0\to([F,F]\cap R)/[F,R]\to R/[F,R]\to F/[F,F]\to G/[G,G]\to 0$.

***Proof* (Sketch) The main point is to show that the obvious surjection**

$H2(BG2,BG1,Z) \to H2(BG,BG1,Z)$

reads out algebraically as the surjection $R/[R,R] \to R/[F,R]$.

In order to explain this, assume that $\{rj|j \in J\}$ is a system of free generators of R. Now, $BG3$ is obtained attaching 3-cells in order to kill the second homotopy group of $BG2$ which, in turn, by Hurewicz' theorem, is the second homology group of its universal cover $\sim BG2$. But it is then easy to see that generators for $H2(\sim BG2,Z)$ are obtained applying the covering transformations of $G=F/R$ to the 2-cells corresponding to the elements $\{rj|j \in J\}$. One can then show that the boundaries of these 3-cells yield relations on $R / [R, R]$ which are exactly those of the form $[a,rj]$, for $a \in F$. ◻

***Example 52* For a finite cyclic group $G=Z/m$, $F=Z$, hence $[F,F]=0$, $H2(Z/m,Z)=0$ and the result fits with previous calculations.**

The spectral sequence argument is however more powerful, and yields a much more general result (see [372] and references therein).

Theorem 53 **(Lyndon–Hochshild–Serre spectral sequence) Let H be any group, and R a normal subgroup, and let $G=H/R$ be the factor group.**

Assume that A is an H-module, so that AH and AH are G-modules. Then there are converging first quadrant spectral sequences

$$E^2_{p,q} = H_p(G; H_q(R, A)) \Rightarrow H_{p+q}(H, A)$$
$$E_2^{p,q} = H^p(G; H^q(R, A)) \Rightarrow H^{p+q}(H, A)$$

***Example 54* Assume that the group H is a metacyclic group, i.e., we have an exact sequence**

$1 \to R \cong Z/n \to H \to G \cong Z/m \to 1.$

As previously remarked, $H2(Z/n,Z)=0$, $H1(Z/n,Z)=Z/n$, while $H0(Z/n,Z)$ is the trivial Z/m-module Z. Hence the spectral sequence has many zeros, and reveals an exact sequence

$Z/m \to H1(Z/m,Z/n) \to H2(H,Z) \to 0.$

Let x be a generator of $R \cong Z/n$, and y a generator of $G \cong Z/m$, which we shall identify to one of its lifts to H(the order of y may be assumed to be equal to m if and only if the exact sequence splits, or, as one says H is**split metacyclic**).

Now, conjugation by y sends x to xa, where $(a,n)=1$, and we have also $am\equiv 1 \ (mod\ n)$. Hence the $G=Z/m$-structure of $R\cong Z/n$, is described by $y(b)=ab\in Z/n$.

Tensoring R with the bar-complex of G we find

$$\oplus_{g_1,g_2\in G}(\mathbb{Z}/n(g_1,g_2)) \to \oplus_{g\in G}(\mathbb{Z}/n)g \to \mathbb{Z}/n$$

where however the last homomorphism is no longer the zero map, and its image is just the image of multiplication by $(a-1)$, i.e.,

$(a-1)(Z/n)=(n,a-1)Z/nZ.$

In fact $Z/n\cong Z[G]\otimes Z[G]Z/n$, with $yi\cdot 1\equiv ai\in Z/n$. Therefore we obtain that $H1(G,Z/n)$ is the quotient of the kernel of multiplication by $(a-1)$, inside the direct sum $\oplus g\in G(Z/n)g$, by the subgroup generated by $g_1 = y^i, g_2 = y^j$.

Writing the elements in G multiplicatively, $g1=yi,g2=yj$, we get the relations

$$y^{i+j} \equiv y^i + a^i y^j.$$

For $i=0$ we get thus $y0=0$, and then inductively we find the equivalent relations

$$y^i = (1 + \cdots + a^{i-1})y, \quad \forall 1 \le i \le m-1,\ 0 = y^m = (1 + \cdots + a^{m-1})y.$$

Hence $H1(G,Z/n)\subset Z/(n,r)$, where we set $r:=(1+\cdots+am-1)= (am-1)/(a-1)$, is the kernel of multiplication by $(a-1)$, hence it is cyclic of order

$$d := \frac{1}{n}(n, a-1)(n, r).$$

The conclusion is that $H2(H,Z)$ is a cokernel of $Z/m\to Z/d$.

Edmonds shows in [146], with a smart trick, that the map $Z/m\to Z/d$ is the zero map when we have a split metacyclic extension, hence $H2(H,Z)\cong Z/d$ in this case. We shall show how to make the computation directly using Hopf's theorem: this has the advantage of writing an explicit generator inside $R / [F, R]$, a fact which shall be proven useful in the sequel.

6.8 Calculating *H2*(*G*,Z) via combinatorial group theory

We treat again the case of a split metacyclic group *H*.

$H=F/R$, where *F* is a free group on two generators *x*, *y*, and the subgroup *R* of relations is normally generated by

$$\xi := x^n, \quad \eta := y^m, \quad \zeta := x^a y x^{-1} y^{-1},$$

with $(a,n)=1$, $am \equiv 1 (\mathrm{mod}\ n)$.

Since *R* is the fundamental group of the Cayley graph of *H*, whose vertices corresponding to the elements of

$$H = \left\{ x^i y^j | 0 \leq i \leq n-1, 0 \leq j \leq m-1 \right\},$$

and *R* is a free group on $mn+1$ generators.

We use now the Reidemeister-Schreier algorithm (see [276]), which shows that a basis of *R* is given by:

$$\eta_i := x^i y^m x^{-i}, \quad \xi := x^n, \quad \xi_{i,j} := x^i y^j x y^{-j} x^{-[i+a^j]},$$
$$0 \leq i \leq n-1, \quad 1 \leq j \leq m-1,$$

and where $[b] \in \{0,...,n-1\}$ is the unique positive representative of the residue class of *b* in $\mathbb{Z}/n$.

We want to calculate $H2(G,\mathbb{Z})$ as the kernel of $R/[F,R] \to K \to 0$, keeping in mind that, since $Fab=F/[F,F]$ is the free abelian group on generators *X*, *Y*, then $K=ker(F_{ab} \to G_{ab})$ is generated by the images of ξ,η,ζ in *Fab*, i.e., $n X$, $m Y$, $(a-1)X$.

We conclude that K is a free abelian group with basis $(n,a-1)X, mY$, and contains the free abelian group with basis $n X$, $m Y$.

Our trick is then to observe that $\xi \mapsto nX$, $\eta = \eta_0 \mapsto mY$, hence $H2(G,\mathbb{Z})$ is the kernel of the surjection

Q:=$(R/[F,R])/(Z\xi+Z\eta))\rightarrow K/(Z(nX)\oplus Z(mY))=(n,a-1)Z/nZ\rightarrow 0$.

This shall simplify our calculations considerably.

Observe that R / $[F, R]$, since $[R,R]\subset[F,R]$, is a quotient of *Rab*. *Rab* is a free abelian group on generators $\Xi,\Xi i,j,\Theta i$, which are the respective images of the elements $\xi,\xi i,j,\eta i$.

Thus R / $[F, R]$, is the quotient of *Rab* by the relations of the form

$$r \sim frf^{-1}, \quad r \in R, \quad f \in F.$$

Down to earth, each generator is put to be equivalent to its conjugate by *x*, respectively its conjugate by *y*.

Conjugation by *x* clearly sends ηi to $\eta i+1$, for all $i<n-1$, while $x\eta_{n-1}x^{-1} = x^n y^m x^{-n} = \xi\eta\xi^{-1}$, thus in the abelianization we get $\Theta i=\Theta,\forall i$, and in our further quotient Q the classes of the Θi's are all zero.

Conjugation of $\xi i,j$ by *x* makes

$$\xi_{i,j} \sim xx^i y^j xy^{-j}x^{-[i+a^j]}x^{-1} = x^{i+1}y^j xy^{-j}x^{-1-[i+a^j]} = \xi_{i+1,j} \quad \text{or} \quad \xi_{i+1,j}\xi^{-1},$$

hence the classes of $\Xi i,j$ and of $\Xi i+1,j$ are the same. Therefore the class of $\Xi i,j$ is always equal to the one of $\Xi 0,j$.

Conjugation by *yj* of ξ=*xn* yields that the class of *yjxny*−*j* is trivial. But, observing that the class of *xn*=ξ and all its conjugates is trivial, we can consider the exponents of powers of *x* as just residue classes modulo *n*, and avoid to use the square brackets. Then

$$y^j x^n y^{-j} = (y^j xy^{-j})^n \sim (y^j xy^{-j}x^{-a^j})(x^{a^j}y^j xy^{-j}x^{-2a^j}) \cdots (x^{(n-1)a^j}y^j xy^{-j}x^{-na^j}),$$

which means that the class of $n\Xi 0,j$=0. We denote by *Zj* the class of $\Xi 0,j$. Hence *nZj*=0, and *Q* is generated by the classes *Zj*.

In the calculation of conjugation by *y* on $\Xi 0,j$ we observe that

$$\xi_{i,j} \sim yy^j xy^{-j}x^{-[a^j]}y^{-1} = (y^{j+1}xy^{-j-1}x^{-[a^{j+1}]})x^{[a^{j+1}]}yx^{-[a^j]}y^{-1}.$$

Hence

$$Z_j \sim yZ_jy^{-1} = Z_{j+1}x^{a^{j+1}}yx^{-a^j}y^{-1} = Z_{j+1}x^{a^{j+1}}yxy^{-1}x^{-a^j-a}x^{a^j+a}yx^{-a^j-1}y^{-1}$$
$$= Z_{j+1}Z_1^k x^{a^{j+1}+ka}yx^{-a^j-k}y^{-1}.$$

For $k=-aj$ we obtained the desired relation

$$Z_j = Z_{j+1}Z_1^{-a^j} \Leftrightarrow Z_{j+1} = Z_jZ_1^{a^j},$$

which can be written additively as

$$y^{i+1} \equiv y^i + a^i y.$$

The conclusion is that we have a cyclic group whose order is $d := \frac{1}{n}(n, a-1)(n, r)$ and with generator $\frac{n}{(n,a-1)}Z_1$, where $Z1$ is the class of $\xi_{0,1} = yxy^{-1}x^{-a}$.

6.9 Sheaves and cohomology on quotients, linearizations

The method of the bar-complex apparently settles the problem of calculating the cohomology groups of a*K*(*G*, 1) space, and also to calculate the cohomology groups *Hn*(*X*,Q) for a rational *K*(*G*, 1) space *X*, except that one must have a good knowledge of the group *G*, for instance a finite presentation does not always suffice.

More generally, Grothendieck [199] approached the question of calculating the cohomology of a quotient *Y*=*X*/*G* where *X* is not necessarily contractible. Using sheaf theory, one can reach a higher generality, considering *G*-linearized sheaves, i.e. sheaves F on *X* with a lifting of the action of *G* to F (this is the analogue to taking cohomology groups on non trivial *G*-modules).

Since this concept is very useful in algebraic geometry, and lies at the heart of some important calculations, we shall try now to briefly explain it, following Mumford's treatment ([303], appendix to section 2, pp. 22–24, section 7, pp. 65–74, section 12, pp. 108–122).

The following is the basic example we have in mind, where we consider two formally different cases:

(1)*X* is a projective variety, *G* is a finite group acting on *X* with the property that for each point *x* the orbit*G x* is contained in an affine open subset of X (this property holds if there is an ample divisor H whose linear equivalence class is fixed by G), or

(2)X is a complex space, G acts properly discontinuously on X (in case where the quotient X / G is a projective variety, we can apply the G.A.G.A. principle by which there is an equivalence of categories between coherent algebraic sheaves and coherent holomorphic sheaves).

Assume that we have a vector bundle V on a quotient variety X / G. Denote by $p{:}X\to X/G$ the quotient map and take the fibre product $p*(V){:=}V\times X/GX$. We consider the action of G on $V\times X$, which is trivial on the factor$V$, and the given one on X. In this way the action of G extends to the pull back $p*(V)=V\times X/GX$ of V, and the sections of V on X / G are exactly seen to be the G-invariant sections of $p*(V)$.

Now, a main goal is to construct interesting varieties as quotient varieties X / G, and then study line bundles on them; the following result is quite useful for this purpose.

Proposition 55 Let $Y=X/G$ be a quotient algebraic variety, $p{:}X\to Y$ the quotient map.

(1)Then there is a functor between (the category of) line bundles L′ on Y and (the category of) G-linearized line bundles L on X, associating to L′ its pull back $p*$(L′). The functor L↦$p*$(L)G is a right inverse to the previous one, and $p*$(L)G is invertible if the action is free, or if both X and Y are smooth.6

(2)Given a line bundle L on X, it admits a G-linearization if and only if there is a Cartier divisor D on Xwhich is G-invariant and such that L≅O$X(D)$ (observe that O$X(D)$={f∈C(X)|$div(f)$+D≥0} has an obvious G-linearization).

(3)A necessary condition for the existence of a linearization is that

$$\forall g \in G, \quad g^*(\mathcal{L}) \cong \mathcal{L}.$$

If this condition holds, defining the Thetagroup of L as:

$$\Theta(\mathcal{L}, G) := \{(\psi, g) | g \in G, \quad \psi : g^*(\mathcal{L}) \cong \mathcal{L} \text{ isomorphism}\},$$

there is an exact sequence

1→C∗→Θ(L,G)→G→1.

The splittings of the above sequence correspond to the G-linearizations of L.

If the sequence splits, the linearizations are a principal homogeneous space over the dual group Hom(G,C∗)=:G∗ of G (namely, each linearization is obtained from a fixed one by multiplying with an arbitrary element in G∗).

Proof **In (1), the case where *G* acts freely is trivial in the holomorphic context, and taken care of by proposition 2, p. 70 of [303] in the algebraic context.**

For (2) and (3) we refer to [302], section 3, chapter 1, and [85], lemmas 4–6.

We only provide an argument for the last statement in (1), for which we do not know of a precise reference. The question is local, hence we may assume that P is a smooth point of X, and that G is equal to the stabilizer of P, hence it is a finite group. The invertible sheaf is locally isomorphic to OX, except that this isomorphism does not respect the linearization. The G-linearization of L corresponds then locally to a character χ:G→C∗. In particular, there exists an integer m such that the induced linearization on L⊗$m$corresponds to the trivial character on OX.

Let R be the ramification divisor of the quotient map p:X→Y=X/G, and let X0:=X\\$Sing$($R$), which is G-invariant. Set Y0:=p(X0).

Step I: p∗(L)G is invertible on Y0.

This is clear on the open set which is the complement of the branch set p(R). At the points of R which are smooth points, then the G action can be linearized locally, and it is a pseudoreflection; so in this case the action of G involves only the last variable, hence the result follows from the dimension 1 case, where every torsion free module is locally free.

Step II. Let M:=p∗(L)G. We claim that M is invertible at every point.

Pick a point y not in Y0 and a local holomorphic Stein neighbourhood U of y, say biholomorphic to a ball. Then$U$0:=U∩Y0 has vanishing homology groups Hi(U0,Z) for i=1,2. By the exponential sequence $H$1(U0,O∗Y)≅$H$1(U0,OY), which is a vector space hence has no torsion elements except the trivial one. SinceM⊗m is trivial on U0, we conclude that M is trivial on U0. By Hartogs' theorem there is then an isomorphism between G-invariant sections of L on p−1(U) and sections of OY on U.

Thus, the question of the existence of a linearization is reduced to the algebraic question of the splitting of the central extension given by the Theta group. This question is addressed by group cohomology theory, as follows (see [223]).

Proposition 56 An exact sequence of groups

$1 \to M \to \Gamma \to G \to 1$,

where M is abelian, is called an extension. Conjugation by lifts of elements of G makes M a G-module. Each choice of a lift $eg \in \Gamma$ for every element $g \in G$ determines a 2-cocycle

$$\psi(g_1, g_2) := e_{g_1} e_{g_2} e^{-1}_{g_1 g_2}.$$

The cohomology class $[\psi] \in H2(G,M)$ is independent of the choice of lifts, and in this way $H2(G,M)$ is in bijection with the set of strict isomorphism classes of extensions $1 \to M \to \Gamma' \to G \to 1$ (i.e., one takes isomorphisms inducing the identity on M and G). Whereas isomorphism classes of extensions $1 \to M' \to \Gamma' \to G \to 1$ (they should only induce the identity on G) are classified by isomorphism classes of pairs (M', ψ').

In particular, we have that the extension is a direct product $M \times G$ if and only if M is a trivial G-module (equivalenty, M is in the centre of Γ) and the class $[\psi] \in H2(G,M)$ is trivial.

Corollary 57 Let L be an invertible sheaf whose class in $Pic(X)$ is G-invariant. Then the necessary and sufficient condition for the existence of a linearization is the triviality of the extension class $[\psi] \in H2(G,C*)$ of the Thetagroup $\Theta(G,L)$.

The group $H2(G,C*)$ is the group of Schur multipliers (see again [223], p. 369) , which for a finite group is, as we saw in Theorem 49, another incarnation of the second homology group $H2(G,Z)$. It occurs naturally when we have a projective representation of a group G. Since, from a homomorphism $G \to PGL(r,C)$ one can pull back the central extension

$1 \to C* \to GL(r,C) \to PGL(r,C) \to 1 \Rightarrow 1 \to C* \to \hat{}G \to G \to 1$,

and the extension class $[\psi] \in H^2(G,C*)$ is the obstruction to lifting the projective representation to a linear representation $G \to GL(r,C)$.

It is an important remark that, if the group G is finite, and $n = ord(G)$, then the cocycles take value in the group of roots of unity $\mu n := \{z \in C* \mid zn = 1\}$.

Example 58 Let E be an elliptic curve, with the point O as the origin,and let G be the group of 2-torsion points $G:=E[2]$acting by translation on E. The divisor class of 2 O is never represented by a G-invariant divisor, since all the$G$-orbits consist of 4 points, and the degree of $2O$ is not divisible by 4. Hence, $L:=O_E(2O)$ does not admit aG-linearization. However, we have a projective representation on $P^1=P(H^0(O_E(2O)))$, where each non zero element η_1 of the group fixes 2 divisors: the sum of the two points corresponding to $\pm\eta_1/2$, and its translate by another point $\eta_2\in E[2]$.

The two group generators yield two linear transformations, which act on $V=\mathbb{C}x_0\oplus\mathbb{C}x_1$ as follows:

$$\eta_1(x_0)=x_1, \eta_1(x_1)=x_0, \eta_2(x_j)=(-1)^j x_j.$$

The linear group generated is however D_4, since

$$\eta_1\eta_2(x_0)=x_1, \eta_1\eta_2(x_1)=-x_0.$$

Example 59 The previous example is indeed a special case of the Heisenberg extension, and V generalizes to the Stone von Neumann representation associated to an abelian group G, which is nothing else than the space $V:=L^2(G,\mathbb{C})$ of square integrable functions on G (see [221, 303]).

G acts on $V=L^2(G,\mathbb{C})$ by translation $f(x)\mapsto f(x-g)$, G^* acts by multiplication with the given character $f(x)\mapsto f(x)\cdot\chi(x))$, and the commutator $[g,\chi]$ acts by the scalar multiplication with the constant $\chi(g)$.

The Heisenberg group is the group of automorphisms of V generated by G,G^* and by $\mathbb{C}^*$ acting by scalar multiplication, and there is a central extension

$$1\to\mathbb{C}^*\to Heis(G)\to G\times G^*\to 1,$$

whose class is classified by the $\mathbb{C}^*$-valued bilinear form $(g,\chi)\mapsto\chi(g)$, an element of

$$\Lambda^2(Hom(G\times G^*,\mathbb{C}^*))\subset H^2(G\times G^*,\mathbb{C}^*).$$

The relation with Abelian varieties $A=V/\Lambda$ is through the Thetagroup associated to an ample divisor L.

In fact, since by the theorem of Frobenius the alternating form $c_1(L)\in H^2(A,\mathbb{Z})\cong\Lambda^2(Hom(\Lambda,\mathbb{Z}))$ admits, in a suitable basis for Λ, the normal form

$$D := \begin{pmatrix} 0 & D' \\ -D' & 0 \end{pmatrix}$$
$$D' := \operatorname{diag}(d_1, d_2, \ldots, d_g), d_1|d_2| \ldots |d_g.$$

The key property is that, if one sets $G:=\mathbb{Z}g/D'\mathbb{Z}g$, then L is invariant exactly for the translations in $(G\times G*)\cong(G\times G)\subset A$, and the Thetagroup of L is just isomorphic to the Heisenberg group $Heis(G)$.

The nice part of the story is the following useful result, which was used by Atiyah in the case of elliptic curves, to study vector bundles over these [12].

Proposition 60 Let G be a finite abelian group, and let $V:=L^2(G,\mathbb{C})$ be the Stone-von Neumann representation. Then $V\otimes V$is a representation of $(G\times G*)$ and splits as the direct sum of all the characters of $(G\times G*)$, taken with multiplicity one.

Proof Since the centre $\mathbb{C}*$ of the Heisenberg group $Heis(G)$ acts trivially on $V\otimes V$, $V\otimes V$ is a representation of $(G\times G*)$. Observe that $(G\times G*)$ is equal to its group of characters, and its cardinality equals to the dimension of $V\otimes V$, hence it shall suffice, and it will be useful for applications, to write for each character an explicit eigenvector.

We shall use the notation g, h, k for elements of G, and the notation χ,η,ξ for elements in the dual group. Observe that V has two bases, one given by $\{g\in G\}$, and the other given by the characters $\{\chi\in G*\}$. In fact, the Fourier transform yields an isomorphism of the vector spaces $V:=L2(G,\mathbb{C})$ and $W:=L2(G*,\mathbb{C})$

$$\mathcal{F}(f) := \hat{f}, \quad \hat{f}(\chi) := \int f(g)(\chi, g)dg.$$

The action of $h\in G$ on V sends $f(g)\mapsto f(g-h)$, hence for the characteristic functions in $\mathbb{C}[G]$, $g\mapsto g+h$. While $\eta\in G*$ sends $f\mapsto f\cdot\eta$, hence $\chi\mapsto\chi+\eta$, where we use the additive notation also for the group of characters.

Restricting V to the finite Heisenberg group which is a central extension of $G\times G*$ by μn, we get a unitary representation, hence we identify V with $\bar{V}$. This said, a basis of $V\otimes\bar{V}$ is given by the set $\{g\otimes\bar{\chi}\}$.

Given a vector $\sum_{g,\chi} a_{g,\chi}(g \otimes \bar{\chi})$the action by $h\in G$ sends it to

$$\sum_{g,\chi}(\chi,h)a_{g-h,\chi}(g\otimes\bar{\chi}),$$

while the one by $\eta\in G*$ sends it to

$$\sum_{g,\chi}(\eta,g)a_{g,\chi-\eta}(g\otimes\bar{\chi}).$$

Hence one verifies right away that

$$F_{k,\xi}:=\sum_{g,\chi}(\chi-\xi,g-k)(g\otimes\bar{\chi})$$

is an eigenvector with eigenvalue $(\xi,h)(\eta,k)$ for $(h,\eta)\in(G\times G*)$. ▯

Remark 61

The explicit calculation reproduced above is based on the fact that the Fourier transform does not commute with the action of the Heisenberg group: the action of G on V corresponds to the action of $G=(G*)*$ on W, while the action of $G*$ on W corresponds to the multiplication of functions $f(g)\in V$ with the conjugate $\bar{\chi}$ of the character χ.

In the next subsection we shall give an explicit example where the Heisenberg group is used for a geometrical construction. Now let us return to the general discussion of cohomology of G-linearized sheaves.

In the special case where the action is free there is an isomorphism between the category of G-linearized sheaves F on X, and sheaves FG on the quotient Y; we have then

Theorem 62 (Grothendieck [199]) Let $Y=X/G$, where G acts freely on X, and let F be a G-linearized sheaf on X, let FG be the G-invariant direct image on Y.

Then there is a spectral sequence converging to a suitable graded quotient of $H^{p+q}(Y,\mathcal{F}^G)$ and with E_1 term equal to $H^p(G,H^q(X,\mathcal{F}))$; we write then as a shorthand notation:

$$H^p(G,H^q(X,\mathcal{F}))\Rightarrow H^{p+q}(Y,\mathcal{F}^G).$$

The underlying idea is simple, once one knows (see [79, 214]) that there is a spectral sequence for the derived functors of a composition of two functors: in fact one takes here the functor $F \mapsto H0(X,F)G$, which is manifestly the composition of two functors $F \mapsto H0(X,F)$ and $A \mapsto AG$, but also the composition of two other functors, since $H0(X,F)G=H0(Y,FG)$.

The main trouble is that the functor $F \mapsto FG$ is no longer exact if we drop the hypothesis that the action is free.

A particular case is of course the one where F is the constant sheaf Z: in the case where the action is not free one has in the calculation to keep track of stabilizers Gx of points $x \in X$, or, even better, to keep track of fixed loci of subgroups $H \subset G$.

Since this bookkeeping is in some situations too difficult or sometimes also of not so much use, Borel [56] proposed (in a way which is similar in spirit to the one of Grothendieck) to consider a generalization of the cohomology of the quotient (see [173] for a nice account of the theory and its applications in algebraic geometry).

Definition 63 Let G be a Lie group acting on a space X, and let $BG=EG/G$ be a classifying space for G. Here we do not need to specify whether G acts on the left or on the right, since for each left action $(g,x) \mapsto gx$ there is the mirror action $xg:=g-1x$.

Letting R be an arbitrary ring of coefficients, the **equivariant cohomology** of X with respect to the action of $G$is defined in one of the following equivalent ways.

(1)Assume that G acts on both spaces X and EG from the left: then we have the diagonal action $g(e,x):=(ge,gx)$ and

$$H_G^i(X, R) := H^i((EG \times X)/G, R).$$

(2)(as in [173]) Assume without loss of generality that G acts on X from the left, and on EG from the right: then

$$H_G^i(X, R) := H^i(EG \times^G X, R),$$

where the space $EG \times GX$ is defined to be the quotient of the Cartesian product $EG \times X$ by the relation

$(e \cdot g,x) \sim (e,g \cdot x)$.

Remark 64 Observe that, in the special case where X is a point, and G is discrete, we reobtain the cohomology group $Hi(G,R)$.

Example 65 The consideration of arbitrary Lie groups is rather natural, for instance if G=C∗, then $BG = \mathbb{P}^{\infty}_{\mathbb{C}} = (\mathbb{C}^{\infty}\setminus\{0\})/\mathbb{C}^*$.

Similarly the infinite Grassmannian $Gr(n$,C∞) is a classifying space for GL(n,C), it is a quotient (cf. [291]) of the Stiefel manifold of frames

$$\mathcal{S}t(n, \mathbb{C}^{\infty}) := \{(v_1, \ldots, v_n) | v_i \in \mathbb{C}^{\infty}, \operatorname{rank}\langle v_1, \ldots, v_n\rangle = n\}.$$

Clearly the universal vector subbundle on the Grassmannian is obtained as the invariant direct image of a GL(n,C)- linearised vector bundle on the Stiefel manifold.

The names change a little bit: instead of a linearized vector bundle one talks here of G-equivariant vector bundles, and there is a theory of equivariant characteristic classes, i.e., these classes are equivariant for morphisms of G-spaces (spaces with a G-action).

Rather than dwelling more on abstract definitions and general properties, I prefer at this stage to mention the most important issues of G-linearized bundles for geometric rational $K(\pi,1)$'s.

6.10 Hodge bundles of weight = 1

(1) Hodge bundle for families of Abelian varieties

Consider the family of Abelian Varieties of dimension g, and with a polarization of type D'~($d1$,$d2$,...,dg)over Siegel's upper half space

$$\mathcal{H}_g := \left\{\tau \in \operatorname{Mat}(g, g, \mathbb{C}) | \tau = {}^t \tau, \operatorname{Im}(\tau) > 0\right\}.$$

The family is given by the quotient of the trivial bundle H:=Hg×Cg by the action of λ∈Z2g acting by

$$(\tau, z) \mapsto (\tau, z + (D', -\tau)\lambda).$$

The local system corresponding to the first cohomology groups of the fibres is the Sp(D,Z)-linearized local system (bundle with fibre Z2g) on HZ:=Hg×Z2g, with the obvious action of Sp(D,Z)

$$M \in \mathrm{Sp}(D, \mathbb{Z}), \quad M(\tau, \lambda) = (M(\tau), M\lambda),$$
$$M := \begin{pmatrix} \alpha & \beta \\ \gamma & \delta \end{pmatrix}, \quad M(\tau) = -D'(D'\alpha - \tau\gamma)^{-1}(D'\beta - \tau\delta).$$

Obviously HZ⊗ZC splits as H1,0⊕H0,1 and the above vector bundle *H* equals *H*1,0 and is called the Hodge bundle.

Stacks are nowadays the language in order to be able to treat the above local system and the Hodge bundle as living over the quotient moduli space A*g*,*D*′:=H*g*/Sp(*D*,Z).

(2) Hodge Bundle for families of curves

Over Teichmüller space we have a universal family of curves, which we shall denote by *pg*:C*g*→T*g*.

The reason is the following: letting *M* be here a compact oriented Riemann surface of genus *g*, and C(*M*) the space of complex structures on *M*, it is clear that there is a universal tautological family of complex structures parametrized by C(*M*), and with total space

$$\mathfrak{U}_{\mathcal{C}(M)} := M \times \mathcal{C}(M),$$

on which the group D*iff*+(*M*) naturally acts, in particular D*iff*0(*M*).

A rather simple consequence of the Lefschetz' Lemma 34 is that D*iff*0(*M*) acts freely on C(*M*): in fact Lefschetz' Lemma 34 implies that for each complex structure *C* on *M* the group of biholomorphisms *Aut*(*C*) contains no automorphism (other than the identity) which is homotopic to the identity, hence a fortiori no one that is differentiably isotopic to the identity.

We have the local system *HZ*:=R¹(*pg*)∗(Z), and again

$$H_{\mathbb{Z}} \otimes_{\mathbb{Z}} \mathbb{C} = H^{1,0} \oplus H^{0,1}, \quad H^{1,0} := (p_g)_* \left(\Omega^1_{\mathcal{C}_g|\mathcal{T}_g}\right).$$

H:=*H*1,0 is called the Hodge bundle, and it is indeed the pull back of the Hodge bundle on A*g* by the Torelli map, or period map, which associates to a complex structure *C* the Jacobian variety Jac(*C*):=*H*1,0(*C*)∨/*H*1(*M*,Z).

The i-th Chern class of the Hodge bundle yield some cohomology classes which Mumford [304, 307] denoted as the *λi*-class (while the notation λ-class is reserved for λ1).

Other classes, which play a crucial role in Mumford's conjecture ([307], see also theorem 233), are the classes K*i*, defined as

$$\mathcal{K}_i := (p_g)_*(K^{i+1}), \quad K := c_1\left(\Omega^1_{\mathcal{C}_g|\mathcal{T}_g}\right).$$

6.11 A surface in a Bagnera–De Franchis threefold

We want to describe here a construction given in joint work with Ingrid Bauer and Frapporti [37] of a surface *S*with an ample canonical divisor *KS*, and with *K*2*S*=6, $p_g := h^0(\mathcal{O}_S(K_S)) = 1,\ q := h^0(\Omega^1_S) = 1.$

Let *A*1 be an elliptic curve, and let *A*2 be an Abelian surface with a line bundle *L*2 yielding a polarization of type (1, 2) (i.e., the elementary divisors for the Chern class of *L*2 are *d*1=1,*d*2=2). Take as *L*1 the line bundle O*A*1(2*O*), and let *L* be the line bundle on *A*′:=*A*1×*A*2 obtained as the exterior tensor product of *L*1 and*L*2, so that

$$H^0(A', L) = H^0(A_1, L_1) \otimes H^0(A_2, L_2).$$

Moreover, we choose the origin in *A*2 so that the space of sections *H*0(*A*2,*L*2) consists only of *even* sections (hence, we shall no longer be free to further change the origin by an arbitrary translation).

We want to take a Bagnera–De Franchis threefold *X*:=*A*/*G*, where *A*=(*A*1×*A*2)/*T*, and *G*≅*T*≅Z/2, and have a surface *S*⊂*X* which is the quotient of a *G*×*T* invariant that $D \in |L|$, so that $S^2 = \frac{1}{4}D^2 = 6.$

We write as usual *A*1=C/Z⊕Z*τ*, and we let *A*2=C2/Λ2, with *λ*1,*λ*2,*λ*3,*λ*4 a basis of Λ2 on which the Chern class of *L*2 is in Frobenius normal form. We let then

$$G := \{Id, g\}, \quad g(a_1 + a_2) := a_1 + \tau/2 - a_2 + \lambda_2/2, \quad \forall a_1 \in A_1, \quad a_2 \in A_2 \tag{66}$$

$$T := (\mathbb{Z}/2)(1/2 + \lambda_4/2) \subset A = (A_1 \times A_2). \tag{67}$$

Now, *G*×*T* surjects onto the group of two torsion points *A*1[2] of the elliptic curve (by the homomorphism associating to an affine transformation its translation vector) and also on the subgroup (Z/2)(*λ*2/2)⊕(Z/2)(*λ*4/2)⊂*A*1[2], and both *H*0(*A*1,*L*1) and *H*0(*A*2,*L*2) are the Stone–von Neumann representation of the Heisenberg group which is a central Z/2 extension of $G \times T$.

By Proposition 60, since in this case (recall the notation $A=V/\Lambda$), we have $V \cong \bar{V}$ (the only roots of unity occurring are just ±1), we conclude that there are exactly 4 divisors in $|L|$, invariant by $a1+a2 \mapsto a1-a2$, and by $a1+a2 \mapsto a1+\tau/2+a2+\lambda 2/2$, and $a1+a2 \mapsto a1+1/2+a2+\lambda 4/2$.

Hence these four divisors descend to give four surfaces $S \subset X$.

This construction is used in [37] to prove the following.

Theorem 68 Let S be a surface of general type with invariants $K2S=6$, $pg=q=1$ such that there exists an unramified double cover $\hat{S} \to S$ with $q(\hat{S})=3$, and such that the Albanese morphism $\hat{\alpha}:\hat{S} \to A$ is birational onto it image$Z$, a divisor in A with $Z3=12$.

Then the canonical model of $\hat{S}$ is isomorphic to Z, and the canonical model of S is isomorphic to $Y=Z/(\mathrm{Z}/2)$, which a divisor in a Bagnera–De Franchis threefold $X:=A/G$, where $A=(A1 \times A2)/T$, $G \cong T \cong \mathrm{Z}/2$, and where the action is as in (66, 67).

These surfaces exist, have an irreducible four dimensional moduli space, and their Albanese map $\alpha:S \to A1=A1/A1[2]$ has general fibre a non hyperelliptic curve of genus $g=3$.

Proof By assumption the Albanese map $\hat{\alpha}:\hat{S} \to A$ is birational onto Z, and we have $K2\hat{S}=12=K2Z$, since by adjunction $OZ(Z)$ is the dualizing sheaf of Z (so Z restricts to the canonical divisor KZ of Z).

We argue similarly to [35], Step 4 of theorem 0.5, p. 31. Denote by W the canonical model of $\hat{S}$, and observe that by adjunction (see loc. cit.) we have $KW=\hat{\alpha}*(KZ)-\mathrm{U}$, where U is an effective Q-Cartier divisor.

We observe now that KZ is ample (by adjunction, since Z is ample) and KW is also ample, hence we have an inequality,

$$12 = K_W^2 = (\hat{\alpha}^*(K_Z) - \mathfrak{U})^2 = K_Z^2 - (\hat{\alpha}^*(K_Z) \cdot \mathfrak{U}) - (K_W \cdot \mathfrak{U}) \geq K_Z^2 = 12,$$

and since both terms are equal to 12, we conclude that U=0, which means that KZ pulls back to KW hence $W$is isomorphic to Z. We have a covering involution $\iota:\hat{S} \to \hat{S}$, such that $S=\hat{S}/\iota$. Since the action of Z/2 is free on $\hat{S}$, Z/2 also acts freely on Z.

Since $Z^3=12$, Z is a divisor of type (1, 1, 2) in A. The covering involution $\iota:\hat{S}\to\hat{S}$ can be lifted to an involution g of A, which we write as an affine transformation $g(a)=\alpha a+\beta$.

We have Abelian subvarieties $A_1=\ker(\alpha-Id)$, $A_2=\ker(\alpha+Id)$, and since the irregularity of S equals 1, A_1 has dimension 1, and A_2 has dimension 2.

We observe preliminarily that g is fixed point free: since otherwise the fixed point locus would be non empty of dimension one, so it would intersect the ample divisor Z, contradicting that $\iota:Z\to Z$ acts freely.

Therefore $Y=Z/\iota$ is a divisor in the Bagnera–De Franchis threefold $X=A/G$, where G is the group of order two generated by g.

We can then write (using the notation introduced in Proposition 16) the Abelian threefold A as A'/T, and since $\beta_1\notin T_1$ we have only two possible cases.

Case (0) $:T=0$.

Case (1) $:T\cong\mathbb{Z}/2$.

We further observe that since the divisor Z is g-invariant, its polarization is α invariant, in particular its Chern class $c\in\wedge^2(\mathrm{Hom}(\Lambda,\mathbb{Z}))$, where $A=V/\Lambda$.

Since $T=\Lambda/(\Lambda_1\oplus\Lambda_2)$, c pulls back to

$$c'\in\wedge^2(\mathrm{Hom}(\Lambda_1\oplus\Lambda_2,\mathbb{Z}))=\wedge^2(\Lambda_1^\vee)\oplus\wedge^2(\Lambda_2^\vee)\oplus(\Lambda_1^\vee)\otimes(\Lambda_2^\vee),$$

and by invariance $c'=(c_1'\oplus c_2')\in\wedge^2(\Lambda_1^\vee)\oplus\wedge^2(\Lambda_2^\vee)$.

So Case 0) bifurcates in the cases:

Case (0-I) c'_1 is of type (1), c'_2 is of type (1, 2).

Case (0-II) c'_1 is of type (2), c'_2 is of type (1, 1).

Both cases can be discarded, since they lead to the same contradiction. Set $D:=Z$: then D is the divisor of zeros on $A=A_1\times A_2$ of a section of a line bundle L which is an exterior tensor product of L_1 and L_2. Since

$$H^0(A,L)=H^0(A_1,L_1)\otimes H^0(A_2,L_2),$$

and $H0(A1,L1)$ has dimension one in case (0-I), while $H0(A2,L2)$ has dimension one in case (0-II), we conclude that D is a reducible divisor, a contradiction, since D is smooth and connected.

In case (1), we denote $A'{:=}A1\times A2$, and we let D be the inverse image of Z inside A'. Again D is smooth and connected, since $\pi1(\hat{}S)$ surjects onto Λ. Now $D2{=}24$, so the Pfaffian of c' equals 4, and there are a priori several possibilities.

Case (1-I) $c'1$ is of type (1).

Case (1-II) $c'2$ is of type (1, 1).

Case (1-III) $c'1$ is of type (2), $c'2$ is of type (1, 2).

Cases (1-I) and (1-II) can be excluded as case (0), since then D would be reducible.

We are then left only with case (1-III), and we may, without loss of generality, assume that $H0(A1,L1){=}H0(A1,OA1(2O))$, and assume that we have chosen the origin so that all the sections of $H0(A2,L2)$ are even.

We have $A{=}A'/T$, and we may write the generator of T as $t1\oplus t2$, and write $g(a1{+}a2){=}(a1{+}\beta1){+}(a2{-}\beta2)$.

By the description of Bagnera–De Franchis varieties given in Sect. 5.1 we have that $t1$ and $\beta1$ are a basis of the group of 2 torsion points of the elliptic curve $A1$.

Now our divisor D, since all sections of $L2$ are even, is $G\times T$ invariant if and only if it is invariant by T and by translation by β.

This condition however implies that translation by $\beta2$ of $L2$ is isomorphic to $L2$, and similarly for $t2$. It follows that $\beta2,t2$ are basis of the kernel $K2$ of the map $\phi L2{:}A2{\rightarrow}Pic0(A2)$, associating to y the tensor product of the translation of $L2$ by y with the inverse of $L1$.

The isomorphism of $G\times T$ with both $K1{:=}A1[2]$ and $K2$ allows to identify both $H0(A1,L1)$ and $H0(A2,L2)$ with the Stone von Neumann representation $L2(T)$: observe in fact that there is only one alternating function $(G\times T){\rightarrow}Z/2$, independent of the chosen basis.

Therefore, there are exactly 4 invariant divisors in the linear system $|L|$.

Explicitly, if $H0(A1,L1)$ has basis $x0,x1$ and $H0(A2,L2)$ has basis $y0,y1$, then the invariant divisors correspond to the four eigenvectors

$$x_0y_0 + x_1y_1, \quad x_0y_0 - x_1y_1, \quad x_0y_1 + x_1y_0, \quad x_0y_1 - x_1y_0.$$

To prove irreducibility of the above family of surfaces, it suffices to show that all the four invariant divisors occur in the same connected family.

To this purpose, we just observe that the monodromy of the family of elliptic curves $E_\tau := \mathbb{C}/(\mathbb{Z} \oplus \mathbb{Z}\tau)$ on the upperhalf plane has the effect that a transformation in $SL(2,\mathbb{Z})$ acts on the subgroup $E_\tau[2]$ of points of 2-torsion by its image matrix in $GL(2,\mathbb{Z}/2)$, and in turn the effect on the Stone von Neumann representation is the one of twisting it by a character of $E_\tau[2]$.

This concludes the proof that the moduli space is irreducible of dimension 4, since the moduli space of elliptic curves, respectively the moduli space of Abelian surfaces with a polarization of type (1, 2), are irreducible, of respective dimensions 1, 3. The final assertion is a consequence of the fact that $Alb(S) = A_1/(T_1 + \langle\langle \beta_1 \rangle\rangle)$, so that the fibres of the Albanese map are just divisors in A_2 of type (1, 2). Their self intersection equals $4 = 2(g-1)$, hence $g = 3$.

In order to establish that the general fibre of the Albanese map is non hyperelliptic, it suffices to prove the following lemma.

Lemma 69 Let A_2 be an Abelian surface, endowed with a divisor L of type (1, 2), so that there is an isogeny of degree two $f: A_2 \to A'$ onto a principally polarised Abelian surface, with kernel generated by a point t of 2-torsion, and such that $L = f^*(\Theta)$. Then the only curves $C \in |L|$ that are hyperelliptic are contained in the pull backs of a translate of Θ by a point of order 2 for a suitable such isogeny $f: A_2 \to A'$. In particular, the general curve $C \in |L|$ is not hyperelliptic.

Proof Let $C \in |L|$, and consider $D := f_*(C) \in |2\Theta|$. There are two cases.

Case (I): $C + t = C$. Then $D = 2B$, where B has genus 2, so that $C = f^*(B)$, hence, since $2B \equiv 2\Theta$, B is a translate of Θ by a point of order 2. There are exactly two such curves, and for them $C \to B$ is étale.

Case (II): $C + t \neq C$. Then $C \to D$ is birational, $f^*(D) = C \cup (C + t)$. Now, $C + t$ is also linearly equivalent to L, hence $C \cap (C + t)$ meet in the 4 base points of the pencil $|L|$. Hence D has two double points and geometric genus equal to 3. These double points are the intersection points of Θ and a translate by a point of order 2, and are points of 2-torsion.

The sections of $H0(OA'(2\Theta))$ are all even and $|2\Theta|$ is the pull-back of the space of hyperplane sections of the Kummer surface K⊂P3, the quotient $K=A'/\{\pm1\}$.

Therefore the image E' of each such curve D lies in the pencil of planes through 2 nodes of K.

E' is a plane quartic, hence E' has geometric genus 1, and we conclude that C admits an involution σ with quotient an elliptic curve E (normalization of E'), and the double cover is branched in 4 points.

Assume that C is hyperelliptic, and denote by h the hyperelliptic involution, which lies in the centre of $Aut(C)$. Hence we have (Z/2)2 acting on C, with quotient P1. We easily see that there are exactly six branch points, two being the branch points of C/h→P1, four being the branch points of E→P1. It follows that there is an étale quotient $C\to B$, where B is the genus 2 curve, double cover of P1 branched on the six points.

Now, the inclusion $C\subset A2$ and the degree 2 map $C\to B$ induces a degree two isogeny $A2\to J(B)$, and C is the pull back of the Theta divisor of $J(B)$, thus it cannot be a general curve.

QED for the lemma.

Definition 70 *Let us call a surface* S *as in Theorem 68 a* ***Sicilian surface with*** *q=pg=1.*

Observe that the fundamental group of S is isomorphic to the fundamental group Γ of X, and that Γ, fitting into the exact sequence

1→Λ→Γ→G=Z/2→1,

is generated by the union of the set $\{g,t\}$, where

$$g(v_1+v_2)=v_1+\tau/2-v_2+\lambda_2/2$$
$$t(v_1+v_2)=v_1+1/2+v_2+\lambda_4/2$$

with the set of translations by the elements of a basis λ1,λ2,λ3,λ4 of Λ2.

It is therefore a semidirect product of Z5=Λ2⊕Zt with the infinite cyclic group generated by g: conjugation by$g$ acts as −1 on Λ2, and it sends $t\mapsto t$−λ4 (hence 2t−λ4 is an eigenvector for the eigenvalue 1).

We shall now give a topological characterization of Sicilian surfaces with q=pg=1, following the lines of [222].

Observe in this respect that X is a $K(\Gamma,1)$ space, so that its cohomology and homology are just group cohomology, respectively homology, of the group Γ.

Corollary 71

A Sicilian surface S with $q=pg=1$ is characterized by the following properties:

(1) $K2S=6$

(2) $\chi(S)=1$

(3) $\pi1(S)\cong\Gamma$, where Γ is as above,

(4) the classifying map $f:S\to X$, where X is the Bagnera–De Franchis threefold which is a classifying space for Γ, has the property that $f*[S]:=Y$ satisfies $Y3=6$.

In particular, any surface homotopically equivalent to a Sicilian surface is a Sicilian surface, and we get a connected component of the moduli space of surfaces of general type which is stable under the action of the absolute Galois group.7

Proof Since $\pi1(S)\cong\Gamma$, first of all $q(S)=1$, hence also $pg(S)=1$. By the same token there is a double étale cover $\hat{}S\to S$ such that $q(\hat{}S)=3$, and the Albanese image of $\hat{}S$, counted with multiplicity, is the inverse image Z of Y, therefore $Z3=12$. From this, it follows that $\hat{}S\to Z$ is birational, since the class of Z is indivisible.

We may now apply the previous theorem in order to obtain the classification.

Observe finally that the condition $(\hat{}\alpha*\hat{}S)3=12$ is not only a topological condition, it is also invariant under Galois autorphisms. ⍰

7 REGULARITY OF CLASSIFYING MAPS AND FUNDAMENTAL GROUPS OF PROJECTIVE VARIETIES

7.1 Harmonic maps

Given a continuous map $f:M \to N$ of differentiable manifolds, we can approximate it, as already partly explained, by a differentiable one, homotopic to the previous one. Indeed, as we already explained, we may assume that $N \subset \mathrm{R}n, M \subset \mathrm{R}m$ and, by a partition of unity argument, that M is an open set in Rh. Convolution approximates then f by a differentiable function $F1$ with values in a tubular neighbourhood $T(N)$ of N, and then the implicit function theorem applied to the normal bundle provides a differentiable retraction $r:T(N) \to N$. Then $F:=r \circ F1$ is the required approximation, and the same retraction provides a homotopy between f and$F$ (the homotopy between f and $F1$ being obvious).

If however M, N are algebraic varieties, and algebraic topology tells us about the existence of a continuous map f as above, we would wish for more regularity, possibly holomorphicity of the homotopic map F.

Now, Wirtinger's theorem (see [306]) characterises complex submanifolds as area minimizing ones, so the first idea is to try to deform a differentiable mapping f until it minimizes some functional.

We may take the Riemannian structure inherited form the chosen embedding, and assume that $(M,gM),(N,gN)$ are Riemannian manifolds.

If we assume that M is compact, then one defines the **Energy** E(f) of the map as the integral:

$$\mathcal{E}(f) := 1/2 \int_M |Df|^2 d\mu_M,$$

where Df is the derivative of the differentiable map f, $d\mu M$ is the volume element on M, and $|Df|$ is just its norm as a differentiable section of a bundle endowed with a metric:

$$Df \in H^0(M, \mathcal{C}^\infty(TM^\vee \otimes f^*(TN))).$$

Remark 72 *(1)*in more concrete terms, the integral in local coordinates has the form

$$\mathcal{E}(f) := 1/2 \int_M \Sigma_{\alpha,\beta,i,j} \left[(g_N)_{\alpha,\beta} \frac{\partial f_\alpha}{\partial x_i} \frac{\partial f_\beta}{\partial x_j} (g_M)^{-1}_{i,j} \right] \sqrt{det(g_{ij})}.$$

(2)Linear algebra shows that, once we identify *TM*, *TN* with their dual bundles via the Riemannian metrics, |*Df*|2=*Tr*((*Df*)V∘*Df*), hence we integrate over *M* the sum of the eigenvalues of the endomorphism (*Df*)V∘*Df*:*TM*→*TM*.

(3)The first variation of the energy function vanishes precisely when *f* is a **harmonic** map, i.e., Δ(*f*)=0, where Δ(*f*):=*Tr*(∇(*Df*)), ∇ being the connection on *TM*V⊗*f*∗(*TN*) induced by the Levi-Civita connections on *M* and *N*.

(4)The energy functional enters also in the study of geodesics and Morse theory (see [288])

These notions were introduced by Eells and Sampson in the seminal paper [147], which used the **heat flow**

$$\frac{\partial f_t}{\partial t} = \Delta(f)$$

in order to find extremals for the energy functional. These curves in the space of maps are (as explained in [147]) the analogue of gradient lines in Morse theory, and the energy functional decreases on these lines.

The obvious advantage of the flow method with respect to discrete convergence procedures ('direct methods of the calculus of variations') is that here it is clear that all the maps are homotopic to each other!8

The next theorem is one of the most important results, first obtained in [147]

Theorem 73 (Eells–Sampson) Let M, N be compact Riemannian manifolds, and assume that the sectional curvature KN of$N$ is semi-negative ($KN \leq 0$): then every continuous map $f0:M \to N$ is homotopic to a harmonic map $f:M \to N$. Moreover the equation $\Delta(f)=0$ implies, in case where M, N are real analytic manifolds, the real analyticity of$f$.

Remark 74 Observe that, in the case where N has strictly negative sectional curvature, Hartmann [209] proved the unicity of the harmonic map in each homotopy class.

Not only the condition about the curvature is necessary for the existence of a harmonic representative in each homotopy class, but moreover it constitutes the main source of connections with the concept of classifying spaces, in view of the classical (see [76, 288]) theorem of Cartan–Hadamard establishing a deep link between curvature and topology.

Theorem 75 (Cartan–Hadamard) Suppose that N is a complete Riemannian manifold, with semi-negative ($KN \leq 0$) sectional curvature: then the universal covering ˜N is diffeomorphic to an Euclidean space, more precisely given any two points there is a unique geodesic joining them.

Remark 76 The reader will notice that the hypotheses of Theorem 73 and of Hartmann's theorem apply naturally to two projective curves M, N of the same genus $g \geq 2$, taken with the metric of constant curvature −1 provided by the uniformization theorem: then one may take for $f0$ a diffeomorphism, and apply the result, obtaining a unique harmonic map f which Samson [330] shows to be also a diffeomorphism. The result obtained is that to the harmonic map f one associates a quadratic differential $\eta_f \in H^0(\Omega_M^{\otimes 2}) = H^0(\mathcal{O}_M(2K))$, and that ηfdetermines the isomorphism class of N. This result constitutes another approach to Teichmüller space Tg.

Thus in complex dimension 1 one cannot hope for a stronger result, to have a holomorphic map rather than just a harmonic one. The surprise comes from the fact that, with suitable assumptions, the hope can be realized in higher dimensions, with a small proviso: given a complex manifold X, one can define the conjugate manifold ¯X as the same differentiable manifold, but where in the decomposition $TX \otimes RC = T(1,0) \oplus T(0,1)$ the roles of $T(1,0)$ and $T(0,1)$ are interchanged (this amounts, in case where X is an algebraic variety, to replacing the defining polynomial equations by polynomials obtained from the previous ones by applying complex conjugation to the coefficients, i.e., replacing each $P(x_0, \ldots, x_N)$ by $\overline{P(\overline{x_0}, \ldots, \overline{x_N})}$).

In this case the identity map, viewed as a map $\iota: X \to \bar{X}$ is no longer holomorphic, but antiholomorphic. Assume now that we have a harmonic map $f: Y \to X$: then also $\iota \circ f$ shall be harmonic, but a theorem implying that f must be holomorphic then necessarily implies that there is a complex isomorphism between X and $\bar{X}$. Unfortunately, this is not the case, as one sees, already in the case of elliptic curves; but then one may restrict the hope to proving that f is either holomorphic or antiholomorphic.

A breakthrough in this direction was obtained by Siu [344] who proved several results, that we shall discuss in the next sections.

7.2 Kähler manifolds and some archetypal theorem

The assumption that a complex manifold X is a Kähler manifold is that there exists a Hermitian metric on the tangent bundle $T(1,0)$ whose associated (1, 1) form ξ is closed. In local coordinates the metric is given by

$$h = \Sigma_{i,j} g_{i,j} dz_i d\bar{z}_j, \quad \text{with} \quad d\xi = 0, \quad \xi := (\Sigma_{i,j} g_{i,j} dz_i \wedge d\bar{z}_j).$$

Hodge theory shows that the cohomology of a compact Kähler manifold X has a Hodge–Kähler decomposition, where $H^{p,q}$ is the space of harmonic forms of type (p, q), which are in particular d-closed (and $d*$-closed):

$$H^m(X, \mathbb{C}) = \oplus_{p,q \geq 0, p+q=m} H^{p,q}, \quad H^{q,p} = \overline{H^{p,q}}, \quad H^{p,q} \cong H^q(X, \Omega^p_X).$$

We give just an elementary application of the above theorem, a characterization of complex tori (see [90, 103,110] for other characterizations)

Theorem 77 Let X be a cKM, i.e., a compact Kähler manifold X, of dimension n. Then X is a complex torus if and only if it has the same integral cohomology algebra of a complex torus, i.e. $H*(X,Z)\cong\Lambda*H1(X,Z)$. Equivalently, if and only if $H*(X,C)\cong\Lambda*H1(X,C)$ and $H2n(X,Z)\cong\Lambda 2nH1(X,Z)$

Proof Since $H2n(X,Z)\cong Z$, it follows that $H1(X,Z)$ is free of rank equal to $2n$, therefore dimC$(H1,0)=n$. We consider then, chosen a base point $x0\in X$, the Albanese map

$$a_X : X \to \mathrm{Alb}(X) := H^0(\Omega^1_X)^\vee / \mathrm{Hom}(H^1(X, \mathbb{Z}), \mathbb{Z}), \quad x \mapsto \int_{x_0}^{x} .$$

Therefore we have a map between X and the complex torus T:=Alb(X), which induces an isomorphism of first cohomology groups, and has degree 1, in view of the isomorphism

$$H^{2n}(X, \mathbb{Z}) \cong \Lambda^{2n}(H^1(X, \mathbb{Z})) \cong H^{2n}(T, \mathbb{Z}).$$

In view of the normality of X, it suffices to show that aX is finite. Let Y be a subvariety of X of dimension m>0mapping to a point: then the cohomology (or homology class, in view of Poincaré duality) class of Y is trivial, since the cohomology algebra of X and T are isomorphic. But since X is Kähler, if ξ is the Kähler form, $\int Y\xi m>0$, a contradiction, since this integral depends only (by the closedness of ξ) on the homology class of Y.□

One can conjecture that a stronger theorem holds, namely

Conjecture 78 Let X be a cKM, i.e., a compact Kähler manifold X, of dimension n. Then X is a complex torus if and only if it has the same rational cohomology algebra of a complex torus, i.e. $H*(X,Q)\cong\Lambda*H1(X,Q)$. Equivalently, if and only if $H*(X,C)\cong\Lambda*H1(X,C)$.

Remark 79 Observe that $H*(X,Q)\cong\Lambda*H1(X,Q)\Leftrightarrow H*(X,C)\cong\Lambda*H1(X,C)$ by virtue of the universal coefficients theorem.

The same argument of the previous Theorem 77 yields that, since $H2n(X,Z)\cong Z$, dimC$(H1,0)=n$ and the Albanese map

$$a_X : X \to A := \mathrm{Alb}(X) := H^0(\Omega^1_X)^\vee / \mathrm{Hom}(H^1(X, \mathbb{Z}), \mathbb{Z}), \ x \mapsto \int_{x_0}^{x}$$

is finite, and it suffices to show that it is unramified (étale), since we have an isomorphism $H^1(X,\mathbb{Z}) \cong H^1(A,\mathbb{Z})$.

One sets therefore R to be the ramification divisor of a_X, and $B = a_{X*}(R)$ the branch divisor.

There are two cases:

Case (i)B is an ample divisor, and A and X are projective.

Case (ii): B is non ample, thus, by a result of Ueno [358] there is a subtorus $A_0 \subset A$ such that B is the pull-back of an ample divisor on A/A_0.

Case (ii)can be reduced, via Ueno's fibration to case (i), which is the crucial one.

Since we have an isomorphism of rational cohomology groups, and Poincaré duality holds, we get that $a_{X*}$and a_*X are, in each degree, isomorphisms with rational coefficients.

The first question is to show that the canonical divisor, which is just the ramification divisor R, is an ample divisor.

Then to observe that the vanishing of the topological Euler Poincaré characteristic $e(X)$ and of the Euler Poincaré characteristics $\chi(\Omega^i_X)=0$ for all i implies (using Riemann Roch?) that the first Chern class of X is zero, thus obtaining a contradiction to the ampleness of K_X.

This step works at least in dimension $n=2$, since then we have the Noether formula $e(X)+K^2_X=12\chi(\mathcal{O}_X)$, hence we obtain $K^2_X=0$.

Added in proof: Will Sawin observed that case i) does not occur for varieties of general type, since for these, by a result of Popa and Schnell, holomorphic 1- forms always have zeros. Instead, Debarre, Jiang and Lahoz showed that case ii) leads to the occurrence of certain iterated torus bundles, so that only a proper modification of our conjecture can be true.

Remark 80 Of course the hypothesis that X is Kähler is crucial: there are several examples, due to Blanchard, Calabi, and Sommese [46–48, 69, 348], of complex manifolds which are diffeomorphic to a complex torus but are not complex tori: indeed KX is not linearly equivalent to a trivial divisor (see [99] for references to the cited papers , and [110] for partial results on the question whether a compact complex manifold with trivial canonical divisor, which is diffeomorphic to a torus, is indeed biholomorphic to a complex torus).

The previous Theorem 77 allows a simple generalization which illustrates well the use of topological methods in moduli theory.

Theorem 81 Let $X=A/G$ be a Generalized Hyperelliptic manifold of complex dimension n and assume that Y is a compact Kähler Manifold satisfying the following properties:

(1)$\pi 1(Y)\cong\pi 1(X)\cong\Gamma$

(2)$H2n(Y,\mathrm{Z})\cong H2n(\Gamma,\mathrm{Z})$ via the natural homomorphism of cohomology groups induced by the continuous map of Y to the classifying space $X=B\Gamma$, associated to the homomorphism $\pi 1(Y)\cong\pi 1(X)\cong\Gamma$.

(3)$Hi(Y,\mathrm{C})\cong Hi(\Gamma,\mathrm{C})$ $\forall i$ via the natural homomorphism analogous to the one defined in (2).

Then

(1)Y is a Generalized Hyperelliptic manifold $Y=A'/G$,

(2)if G is abelian, then Y is (real) affinely equivalent to X, and moreover

(3)Y is a complex deformation of X if and only if X and Y have the same Hodge type.

Proof Without loss of generality, we may assume that G contains no translations.

Since $\Gamma=\pi 1(X)$, we have an exact sequence $1\rightarrow\Lambda\rightarrow\Gamma\rightarrow G\rightarrow 0$, and we let W be the unramified covering of Y corresponding to the surjection $\pi 1(Y)\cong\Gamma\rightarrow G$.

Then the Albanese variety A' of W has dimension n, and the group G acts on A'.

Main claim: G acts freely on A'.

This follows since, otherwise, there is an element $g' \in G \setminus 1G$ and a lift g of g' in Γ such that a positive power gm of g equals the neutral element 1Γ (see Step I of the proof of Proposition 21). But then g' would not act freely on A.

Now the Albanese map $W \to A'$ induces a holomorphic map $f: Y \to A'/G$, and we know that A'/G is a classifying space with $\pi 1(A'/G) \cong \Gamma$. By our hypothesis f induces an isomorphism of cohomology groups and has degree equal to 1. The same argument as in Theorem 77 shows that f is an isomorphism. Therefore Y is also a Generalized Hyperelliptic manifold of complex dimension n.

For the second assertion, we simply apply Proposition 21, stating that the exact sequence $1 \to \Lambda \to \Gamma \to G \to 0$ determines, when G is abelian, the real affine type of the action of Γ.

The last assertion is a direct consequence of Remark 20. ▯

The following is one more characterization of complex tori and Abelian varieties.

Corollary 82 Let Z be a projective variety. Then Z is a projective $K(\pi,1)$ with π an abelian group ⇔ Z is an Abelian variety. Similarly, if Z is a cKM, Z is a $K(\pi,1)$ with π abelian ⇔ Z is a complex torus.

Proof Since π is finitely generated, we can write $\pi = \mathbb{Z}m \oplus T$, where T is a finite abelian group. The subgroup Zm is the fundamental group of a Galois cover $W \to Z$ with group T. In this case m must be even $m=2n$, by the Kähler assumption, and since $H*(W,\mathbb{Z}) = H*(\mathbb{Z}2n,\mathbb{Z})$, we obtain that $n = dim\mathbb{C}(W) = dim\mathbb{C}(Z)$ and we can apply Theorem 77 to infer that W is a complex torus.

Now, $Z = W/T$, where T is abelian, and T acts trivially by conjugation on $\pi 1(W) \cong \mathbb{Z}2n$. Hence T is a group of translations, and we obtain that Z is a complex torus. Moreover, since the fundamental group of a torus is torsion free, we can actually conclude that $T=0$.

7.3 Siu's results on harmonic maps

The result by Siu that is the simplest to state is the following

Theorem 83 (I) Assume that $f{:}M\to N$ is a harmonic map between two compact Kähler manifolds and that the curvature tensor of N is strongly negative. Assume further that the real rank of the derivative Df is at least 4 in some point of M. Then f is either holomorphic or antiholomorphic.

(II) In particular, if $\dim_{\mathbb{C}}(N)\geq 2$ and M is homotopy equivalent to N, then M is either biholomorphic or antibiholomorphic to N.

From Theorem 83 follows an important consequence.

Corollary 84 Assume that $f{:}M\to N$ is a continuous map between two compact Kähler manifolds and that the curvature tensor of N is strongly negative. Assume further that there is a $j\geq 4$ such that $Hj(f,\mathbb{Z})\neq 0$: then f is homotopic to a map F which is either holomorphic or antiholomorphic.

Proof f is homotopic to a harmonic map F. One needs to show that at some point the real rank of the derivative $DF$is at least 4. If it were not so, then by Sard's lemma the image $Y{:=}F(M)$ would be a subanalytic compact set of Hausdorff dimension at most 3.

Lemma 85 Let Y be a subanalytic compact set, or just a compact set of Hausdorff dimension at most h. Then $Hi(Y,\mathbb{Z})=0$for $i\geq h+1$.

The lemma then implies $Hi(Y,\mathbb{Z})=0$, $\forall i\geq 4$, contradicting the existence of an $i\geq 4$ such that $Hi(f,\mathbb{Z})\neq 0$.

First proof of the lemma: Y is a finite union of locally closed submanifolds of dimension $\leq h$, by the results of [42]. Using the exact sequence of Borel-Moore homology (see [59]) relating the homology of $Y=U\cup F$, where Y is the union of a closed set F with an open subset U, and induction on the number of such locally closed sets, we obtain that $Hi(Y,\mathbb{Z})=0$ for $i\geq h+1$.

Second proof of the lemma: Y is a compact of Hausdorff dimension at most h. By theorem VII 3 of [219], page 104, follows that the dimension of Y is at most h. In turn, using theorem 2' , p. 362 of [1], it follows (take Aibidem to be a point P) that $Hi(Y,\mathbb{Z})=Hi(Y,P,\mathbb{Z})=0$ for $i\geq h+1$.

Remark 86 (1) The hypothesis that the rank of the differential should be at least 4 at some point is needed for instance to avoid that the image of F is a complex projective curve C contained inside a projective variety N with such a strongly negative Kähler metric. Because in this case f could be the composition of a holomorphic map $g{:}M\to C'$ to a curve C' of the same genus as C, but not isomorphic to C, composed with a diffeomorphism between C' and C.

(2) Part II of Theorem 83 follows clearly from part I: because, as we have seen, the homotopy equivalence between M and N is realised by a harmonic map $f{:}M\to N$. From part I of the theorem one knows that f is holomorphic or antiholomorphic (in short, one says that f is **dianalytic**). W.l.o.g. let us assume that f is holomorphic (replacing possibly N by $\bar{N}$). There remains only to prove that f is biholomorphic. The argument is almost the same as the one used in the archetypal Theorem 77.

Step 1: f is a finite map; otherwise, since f is proper, there would be a complex curve C such that $f(C)$ is a point. But in a Kähler manifold the homology class in dimension $2m$ of a subvariety of complex dimension m, here C, is never trivial. On the other hand, since f is a homotopy equivalence, and $f(C)$ is a point, this class must be zero.

Step 2: f is a map of degree one since f induces an isomorphism between the last non zero homology groups of M, N respectively. If $n=\dim_{\mathbb{C}} N, m=\dim_{\mathbb{C}} M$, then these groups are $H_{2m}(M,\mathbb{Z})$, respectively $H_{2n}(N,\mathbb{Z})$; hence $n=m$ and

$$H_{2m}(f) : H_{2m}(M,\mathbb{Z}) \cong \mathbb{Z}[M] \to H_{2m}(N,\mathbb{Z}) \cong \mathbb{Z}[N]$$

is an isomorphism.

Step 3: f is holomorphic, finite and of degree 1.Therefore there are open subsets whose complements are Zariski closed such that $f{:}U\to V\subset N$ is an isomorphism. Then the inverse of f is defined on V, the complement of a complex analytic set, and by the Riemann extension theorem (normality of smooth varieties) the inverse extends to N, showing that f is biholomorphic.

(3) We have sketched the above argument since it appears over and over in the application of homotopy equivalence to proving isomorphism of complex projective (or just Kähler) manifolds. Instead, the proof of part I is based on the Bochner–Nakano formula, which was later further generalised by Siu [346], and is too technical to be fully discussed here.

Let us try however to describe precisely the main hypothesis of strong negativity of the curvature, which is a stronger condition than the strict negativity of the sectional curvature.

As we already mentioned, the assumption that N is a Kähler manifold is that there exists a Hermitian metric on the tangent bundle $T(1,0)$ whose associated (1, 1) form is closed. In local coordinates the metric is given by

$$\Sigma_{i,j} g_{i,j} dz_i d\bar{z}_j, \quad \text{with} \quad d(\Sigma_{i,j} g_{i,j} dz_i \wedge d\bar{z}_j) = 0.$$

The curvature tensor is a (1, 1) form with values in $(T^{(1,0)})^\vee \otimes T^{(1,0)}$, and using the Hermitian metric to identify $(T^{(1,0)})^\vee \cong \overline{T^{(1,0)}} = T^{(0,1)}$, and their conjugates $((T^{(0,1)})^\vee = \overline{(T^{(0,1)})} \cong T^{(1,0)})$ we write as usual the curvature tensor as a section R of

$$(T^{(1,0)})^\vee \otimes (T^{(0,1)})^\vee \otimes (T^{(1,0)})^\vee \otimes (T^{(0,1)})^\vee.$$

Then seminegativity of the sectional curvature is equivalent to

$$-R\left(\xi \wedge \bar{\eta} - \eta \wedge \bar{\xi}, \overline{\xi \wedge \bar{\eta} - \eta \wedge \bar{\xi}}\right) \leq 0,$$

for all pairs of complex tangent vectors ξ,η (here one uses the isomorphism $T(1,0)\cong TN$, and one sees that the expression depends only on the real span of the two vectors ξ,η).

Strong negativity means instead that

$$-R\left(\xi \wedge \bar{\eta} - \zeta \wedge \bar{\theta}, \overline{\xi \wedge \bar{\eta} - \zeta \wedge \bar{\theta}}\right) < 0,$$

for all 4-tuples of complex tangent vectors ξ,η,ζ,θ.

The geometrical meaning is the following (see [2], p. 71): the sectional curvature is a quadratic form on $\Lambda 2(TN)$, and as such it extends to the complexified bundle $\Lambda 2(TN)\otimes \mathbb{C}$ as a Hermitian form. Then strong negativity in the sense of Siu is also called negativity of the **Hermitian sectional curvature** $R(v,w,\bar{v},\bar{w})$ for all vectors $v,w\in(TN)\otimes \mathbb{C}$.

Then a reformulation of the result of Siu [344] and Sampson [331] is the following:

Theorem 87 Let *M* be a compact Kähler manifold, and *N* a Riemannian manifold with semi-negative Hermitian sectional curvature. Then every harmonic map *f*:*M*→*N* is pluri-harmonic.

Now, examples of varieties *N* with a strongly negative curvature are the balls in C*n*, i.e., the BSD of type *In*,1; Siu finds out that ([344], see also [70]) for the irreducible bounded symmetric domains of type

$$I_{p,q}, \quad \text{for } pq \geq 2,\ II_n, \quad \forall n \geq 3, III_n, \quad \forall n \geq 2, IV_n, \quad \forall n \geq 3,$$

the metric is not strongly negative, but just very strongly seminegative, where very strong negativity simply means negativity of the curvature as a Hermitian form on $T^{1,0} \otimes T^{0,1} = T^{1,0} \otimes \overline{T^{0,1}}$..

Indeed, the bulk of the calculations is to see that there is an upper bound for the nullity of the Hermitian sectional curvature, i.e. for the rank of the real subbundles of *TM* where the Hermitian sectional curvature restricts identically to zero (in Siu's notation, then one considers always the case where very strongly seminegativity holds, and 2-negative means strongly negative, adequately negative means that the nullity cannot be maximal).

Hence Siu derives several results,

Theorem 88 (Siu) Suppose that *M*, *N* are compact Kähler manifolds and the curvature tensor of *N* is negative of order *k*. Assume that *f*:*M*→*N* is a harmonic map such that, ∃*i*≥2*k* such that *Hi*(*f*,Z)≠0. Then *f* is either holomorphic or antiholomorphic.

Theorem 89 (Siu) Suppose that *M* is a compact Kähler manifold and *N* is a locally symmetric manifold D/Γ, where D is an irreducible bounded symmetric domain of one of the following types:

$$I_{p,q}, \quad \text{for} \quad pq \geq 2,\ II_n, \quad \forall n \geq 3, III_n, \quad \forall n \geq 2, IV_n, \quad \forall n \geq 3.$$

Assume that *f*:*M*→*N* is a harmonic map such that, *n*:=*dim*C*N*, *H*2*n*(*f*,Z)≠0. Then *f* is either holomorphic or antiholomorphic.

In particular, if *M* is a compact Kähler manifold homotopically equivalent to *N* as above, then either *M*≅*N* or *M*≅¯*N*.

The most general result is ([2], p. 80, theorems 6.13–15):

Theorem 90 (Siu) Suppose that M is a compact Kähler manifold and N is a locally Hermitian symmetric space $D/\Gamma=G/K$, where the irreducible decomposition of D contains no dimension 1 factors.

(1)Assume that $f:M\to N$ is a harmonic map such that, $n:=dim\mathbb{C}N$, $H2n(f,\mathbb{Z})\neq 0$. Then f is holomorphic for some invariant complex structure on G / K.

(2)Assume that M is homotopically equivalent to N: then f is biholomorphic to G / K for some invariant complex structure on G / K.

*Remark 91*A few words are needed to explain the formulation 'for some invariant complex structure on G / K'. In the case where D is irreducible, we can just take the conjugate complex structure. But if $D=D1\times\cdots\times Dl$, one can just take the conjugate complex structure on a subset I of the indices $j\in\{1,\ldots,l\}$.

7.4 Hodge theory and existence of maps to curves

Siu also used harmonic theory in order to construct holomorphic maps from Kähler manifolds to projective curves. The first result in this direction was the theorem of [347], also obtained by Jost and Yau (see [239] and also [228] for other results).

Theorem 92 (Siu) Assume that a compact Kähler manifold X is such that there is a surjection $\phi:\pi 1(X)\to\pi g$, where $g\geq 2$and, as usual, πg is the fundamental group of a projective curve of genus g. Then there is a projective curve $C$of genus $g'\geq g$ and a fibration $f:X\to C$ (i.e., the fibres of f are connected) such that ϕ factors through $\pi 1(f)$.

In this case the homomorphism leads to a harmonic map to a curve, and one has to show that the Stein factorization yields a map to some Riemann surface which is holomorphic for some complex structure on the target.

In this case it can be seen more directly how the Kähler assumption, which boils down to Kähler identities, is used.

Recall that Hodge theory shows that the cohomology of a compact Kähler manifold X has a Hodge–Kähler decomposition, where Hp,q is the space of harmonic forms of type (p, q):

$$H^m(X,\mathbb{C}) = \oplus_{p,q\geq 0,\, p+q=m} H^{p,q}, \quad H^{q,p} = \overline{H^{p,q}}, \quad H^{p,q} \cong H^q(X,\Omega_X^p).$$

The Hodge–Kähler decomposition theorem has a long story, revived by Griffths in [192, 193], and was proven by Picard in special cases. It entails the following consequence:

Holomorphic forms are closed, i.e., $\eta \in H^0(X,\Omega^p_X) \Rightarrow d\eta = 0$.

At the turn of last century this fact was then used by Castelnuovo and de Franchis [80, 130]:

Theorem 93 (Castelnuovo–de Franchis) Assume that X is a compact Kähler manifold, $\eta_1, \eta_2 \in H^0(X, \Omega^1_X)$ are C-linearly independent, and the wedge product $\eta_1 \wedge \eta_2$ is d-exact. Then $\eta_1 \wedge \eta_2 \equiv 0$ and there exists a fibration $f: X \to C$ such that $\eta_1, \eta_2 \in f^* H^0(C, \Omega^1_C)$ In particular, C has genus $g \geq 2$.

Even if the proof is well known, let us point out that the first assertion follows from the Hodge–Kähler decomposition, while $\eta_1 \wedge \eta_2 \equiv 0$ implies the existence of a non constant rational function φ such that $\eta_2 = \varphi \eta_1$. This shows that the foliation defined by the two holomorphic forms has Zariski closed leaves, and the rest follows then rather directly taking the Stein factorization of the rational map $\varphi: X \to \mathbb{P}^1$.

Now, the above result, which is holomorphic in nature, combined with the Hodge decomposition, produces results which are topological in nature (they actually only depend on the cohomology algebra structure of $H^*(X, \mathbb{C})$).

To explain this in the most elementary case, we start from the following simple observation. If two linear independent vectors in the first cohomology group $H^1(X, \mathbb{C})$ of a Kähler manifold have wedge product which is trivial in cohomology, and we represent them as $\eta_1 + \bar{\omega}_1, \eta_2 + \bar{\omega}_2$, for $\eta_1, \eta_2, \omega_1, \omega_2 \in H^0(X, \Omega^1_X)$, then by the Hodge decomposition and the first assertion of the theorem of Castelnuovo–de Franchis

$$(\eta_1 + \overline{\omega_1}) \wedge (\eta_2 + \overline{\omega_2}) = 0 \in H^2(X, \mathbb{C})$$

implies

$$\eta_1 \wedge \eta_2 \equiv 0, \quad \omega_1 \wedge \omega_2 \equiv 0.$$

We can apply Castelnuovo–de Franchis unless $\eta 1,\eta 2$ are C-linearly dependent, and similarly $\omega 1,\omega 2$. W.l.o.g. we may assume $\eta 2\equiv 0$ and $\omega 1\equiv 0$. But then $\eta 1\wedge\bar{}\omega 2=0$ implies that the Hodge norm

$$\int_X (\eta_1 \wedge \overline{\omega_2}) \wedge \overline{(\eta_1 \wedge \overline{\omega_2})} \wedge \xi^{n-2} = 0,$$

where ξ is here the Kähler form. A simple trick is to observe that

$$0 = \int_X (\eta_1 \wedge \overline{\omega_2}) \wedge \overline{(\eta_1 \wedge \overline{\omega_2})} \wedge \xi^{n-2} = -\int_X (\eta_1 \wedge \omega_2) \wedge \overline{(\eta_1 \wedge \omega_2)} \wedge \xi^{n-2},$$

therefore the same integral yields that the Hodge norm of $\eta 1\wedge\omega 2$ is zero, hence $\eta 1\wedge\omega 2\equiv 0$; the final conclusion is that we can in any case apply Castelnuovo-de Franchis and find a map to a projective curve $C$of genus $g\geq 2$.

More precisely, one gets the following theorem [87]:

Theorem 94 (Isotropic subspace theorem) On a compact Kähler manifold X there is a bijection between isomorphism classes of fibrations $f:X\to C$ to a projective curve of genus $g\geq 2$, and real subspaces $V\subset H1(X,\mathrm{C})$ ('real' means that V is self conjugate, $\bar{}V=V$) which have dimension $2g$ and are of the form $V=U\oplus\bar{}U$, where U is a maximal isotropic subspace for the wedge product

$$H^1(X, \mathbb{C}) \times H^1(X, \mathbb{C}) \to H^2(X, \mathbb{C}).$$

It is interesting that the above result implies the following theorem of Gromov [196], which in turn obviously implies theorem 92 of Siu (see [25, 197] for related results).

Theorem 95

(Gromov's few relations theorem) Let X be a compact Kähler manifold and assume that there exists a surjection of its fundamental group

$\Gamma:=\pi 1(X)\to G=\langle x1,\ldots,xn|R1(x),\ldots,Rm(x)\rangle$,

onto a finitely presented group that has 'few relations', more precisely where $n\geq m-2$. Then there exists a fibration $f:X\to C$, onto a projective curve C of genus $g \geq \frac{1}{2}(n-m)$.

If moreover $\pi 1(X)\cong G$, then the first cohomology group $H1(X,\mathrm{C})$ equals $f*H1(C,\mathrm{C})$.

Proof We saw in a Sect. 6.6 that if the fundamental group of X, $\Gamma:=\pi 1(X)$ admits a surjection onto G, then the induced classifying continuous map $\phi:X\to BG$ has the properties that its induced action on first cohomology

$$H^1(\phi) : H^1(BG, \mathbb{C}) \to H^1(X, \mathbb{C})$$

is injective and the image W of $H1(\phi)$ is such that each element $w\in W$ is contained in an isotropic subspace of rank $\geq n-m$.

Hence it follows immediately that there is an isotropic subspace of dimension $\geq n-m$ and a fibration onto a curve of genus $g\geq n-m$.

For the second assertion, let U be a maximal isotropic subspace contained in $W=H1(X,\mathrm{C})$: then there exists a fibration $f:X\to C$, onto a projective curve C of genus $g\geq 2$ such that $f*H1(C,\mathrm{C})=U\oplus^{-}U$. We are done unless $U\oplus^{-}U\neq H1(X,\mathrm{C})$.

But in this case we know that $W=H1(X,\mathrm{C})$ is the union of such proper subspaces $U\oplus^{-}U$.

These however are of the form $f*H1(C,\mathrm{Z})\otimes\mathrm{C}$, hence they are a countable number; by Baire's theorem their union cannot be the whole C-vector space $W=H1\ (X,\mathrm{C})$. ▯

Not only one sees clearly how the Kähler hypothesis is used, but indeed Kato [233] and Pontecorvo [324] showed how the results are indeed false without the Kähler assumption, using twistor spaces of algebraic surfaces which are P1-bundles over a projective curve C of genus $g\geq 2$. A simpler example was then found by Kotschick ([2], ex. 2.16, p. 28): a primary Kodaira surface (an elliptic bundle over an elliptic curve, with $b1(X)=3, b2(X)=4$).

We do not mention in detail generalisations of the Castelnuovo–de Franchis theory to higher dimensional targets (see [87, 190, 191, 342]), since these shall not be used in the sequel.

We want however to mention another result ([106], see also [102] for a weaker result) which again, like the isotropic subspace theorem, determines explicitly the genus of the target curve (a result which is clearly useful for classification and moduli problems).

Theorem 96 Let X be a compact Kähler manifold, and let $f:X\to C$ be a fibration onto a projective curve C, of genus g, and assume that there are exactly r fibres which are multiple with multiplicities $m1,\ldots mr\geq 2$. Then f induces an orbifold fundamental group exact sequence

$$\pi 1(F)\to\pi 1(X)\to\pi 1(g;m1,\ldots mr)\to 0,$$

where F is a smooth fibre of f, and

$$\pi_1(g;m_1,\ldots m_r)$$
$$:= \langle\alpha_1,\beta_1,\ldots,\alpha_g,\beta_g,\gamma_1,\ldots\gamma_r|\ \Pi_1^g[\alpha_j,\beta_j]\Pi_1^r\gamma_i=\gamma_1^{m_1}=\cdots=\gamma_r^{m_r}=1\rangle.$$

Conversely, let X be a compact Kähler manifold and let $(g,m1,\ldots mr)$ be a hyperbolic type, i.e., assume that $2g-2+\Sigma_i(1-\frac{1}{m_i})>0.$.

Then each epimorphism $\phi:\pi 1(X)\to\pi 1(g;m1,\ldots mr)$ with finitely generated kernel is obtained from a fibration $f:X\to C$ of type $(g;m1,\ldots mr)$.

7.5 Restrictions on fundamental groups of projective varieties

A very interesting (and still largely unanswered) question posed by Serre [335] is:

Question 97 (Serre)

(1)Which are the **projective groups,** i.e., the groups π which occur as fundamental groups $\pi=\pi 1(X)$ of a complex projective manifold?

(2)Which are the **Kähler groups,** i.e., the groups π which occur as fundamental groups $\pi=\pi 1(X)$ of a compact Kähler manifold?

Remark 98 *(i)*Serre himself proved (see [337]) that the answer to the first question is positive for every finite group. In this chapter we shall only limit ourselves to mention some examples and results to give a general idea, especially about the use of harmonic maps, referring the reader to the book [2], entirely dedicated to this subject.

(ii)For question (1), in view of the Lefschetz hyperplane section Theorem 3, the class of projective groupsπ is exactly the class of groups which occur as fundamental groups $\pi=\pi 1(X)$ of a complex projective smooth surface.

(iii) if π and π' are projective (resp. Kähler), the same is true for the Cartesian product (take $X \times X'$).

The first obvious restriction for a group Γ to be a Kähler group (only a priori the class of Kähler groups is a larger class than the one of projective groups) is that their first Betti number $b1$ (rank of the abelianization $\Gamma ab = \Gamma/[\Gamma,\Gamma]$) is even; since if $\Gamma = \pi 1(X)$, then $\Gamma ab = H1(X,Z)$, and $H1(X,Z) \otimes C$ has even dimension by the Hodge–Kähler decomposition.

A more general restriction is that the fundamental group Γ of a compact differentiable manifold M must be finitely presented: since by Morse theory (see [288]) M is homotopically equivalent to a finite CW-complex ($M$is obtained attaching finitely many cells of dimension $\le \dim R(M)$).

Conversely (see [333], p. 180),

Theorem 99 Any finitely presented group Γ is the fundamental group of a compact oriented 4-manifold.

Recall for this the

Definition 100 (*Connected sum*) Given two differentiable manifolds $M1,M2$ of the same dimension n, take respective points $Pi \in Mi, i=1,2$ and respective open neighbourhoods Bi which are diffeomorphic to balls, and with smooth boundary $\partial Bi \cong Sn-1$. Glueing together the two manifolds with boundary $Mi \setminus Bi, \partial Bi$ we obtain a manifold $M1 \sharp M2$, which is denoted by the connected sum of $M1,M2$ and whose diffeomorphism class is independent of the choices made in the construction. Also, if $M1,M2$ are oriented, the same holds for $M1 \sharp M2$, provided the diffeomorphisms $\partial Bi \cong Sn-1$ are chosen to be orientation preserving.

Definition 101

(*Free product*) If Γ,Γ' are finitely presented groups,

$\Gamma = \langle x1,\ldots,xn \mid R1(x),\ldots,Rm(x)\rangle, \Gamma' = \langle y1,\ldots,yh \mid R'1(y),\ldots,R'k(y)\rangle,$

their free product is the finitely presented group

$\Gamma * \Gamma' = \langle x1,\ldots,xn,y1,\ldots,yh \mid R1(x),\ldots,Rm(x),R'1(y),\ldots,R'k(y)\rangle.$

Idea for the proof of theorem 99: By the van Kampen theorem, the fundamental group of the union $X = X1 \cup X2$ of two open sets with connected intersection $X1 \cap X2$ is the quotient of the free product

$\pi 1(X1)*\pi 1(X2)$ by the relations $i1(\gamma)=i2(\gamma),\forall\gamma\in\pi 1(X1\cap X2)$, where $ij:\pi 1(X1\cap X2)\to\pi 1(Xj)$ is induced by the natural inclusion $(X1\cap X2)\subset Xj$.

Now, assume that Γ has a finite presentation

$\Gamma=\langle x1,\ldots,xn|R1(x),\ldots,Rm(x)\rangle$.

Then one considers the connected sum of n copies of $S1\times S3$, which has then fundamental group equal to a free group $\mathrm{F}n=\langle\langle x1,\ldots,xn\rangle\rangle$.

Realizing the relations $Rj(x)$ as loops connecting the base point to m disjoint embedded circles (i.e., diffeomorphic to $S1$), one can perform the so called **surgery** replacing tubular neighbourhoods of these circles, which are diffeomorphic to $S1\times B3$, and have boundary $S1\times S2$, each by the manifold with boundary $B2\times S2$. These manifolds are simply connected, hence by van Kampen we introduce the relation $Rj(x)=1$, and finally one obtains a M with fundamental group $\cong\Gamma$.

There are no more restrictions, other than finite presentability, if one requires that Γ be the fundamental group of a compact complex manifold, as shown by Taubes [351], or if one requires that Γ be the fundamental group of a compact symplectic 4-manifold, as shown by Gompf [182]. Notice however that, by the main results of Kodaira's surface classification (see [21]), the fundamental groups of complex non projective surfaces form a very restricted class, in particular either their Betti number $b1$ is equal to 1, or they sit in an exact sequence of the form

$1\to\mathrm{Z}\to\Gamma\to\pi g\to 1$.

In fact Taubes builds on the method of Seifert and Threlfall in order to construct a compact complex manifold$X$ with $\pi 1(X)=\Gamma$. He takes a (differentiable) 4-manifold M with $\pi 1(M)=\Gamma$ and then, taking the connected sum with a suitable number of copies of the complex projective plane with opposed orientation, he achieves that Mallows a metric with anti self dual Weyl tensor. This condition on the metric of M, by a theorem of Atiyah, Hitchin and Singer [14] makes the twistor space $Tw(M)$ a complex manifold (not just an almost complex manifold). Note that (see [268], especially page 366) the twistor space $Tw(M)$ of an oriented Riemannian 4-manifold is an $S2$-bundle over M, such that the fibre over $P\in M$ equals the sphere bundle of the rank 3 vector subbundle $\Lambda+\subset\Lambda 2(TM)$. Here $\Lambda 2(TM)=\Lambda+\oplus\Lambda-$ is the eigenspace decomposition for the $*$-operator $*:\Lambda 2(TM)\to\Lambda 2(TM)$, such that $*2=1$.

The fact that $Tw(M)$ is an $S2$-bundle over M is of course responsible of the isomorphism $\pi 1(Tw(M)) \cong \pi 1(M)$.

Returning to Serre's question, a first result was given by Johnson and Rees [225], which was later extended by other authors [7, 196]:

Theorem 102 (Johnson–Rees) Let $\Gamma 1, \Gamma 2$ be two finitely presented groups admitting some non trivial finite quotient. Then the free product $\Gamma 1 * \Gamma 2$ cannot be the fundamental group of a normal projective variety. More generally, this holds for any direct product $H \times (\Gamma 1 * \Gamma 2)$.

The idea of proof is to use the fact that the first cohomology group $H1(X,\mathrm{C})$ carries a nondegenerate skew-symmetric form, obtained simply from the cup product in 1-dimensional cohomology multiplied with the $(n-1)$-th power of the Kähler class.

An obvious observation is that the crucial hypothesis is projectivity. In fact, if we take a quasi-projective variety, then the theorem does not hold true: it suffices to take as X the complement of d points in a projective curve of genus g, and then $\pi 1(X) \cong \mathrm{F}2g+d-1$ is a free group. This is not just a curve phenomenon: already Zariski showed that if C is a plane sextic with equation $Q2(x)3-G3(X)2=0$, where $Q2,G3$ are generic forms of respective degrees 2, 3, then $\pi 1(\mathrm{P}2 \backslash C) \cong \mathrm{Z}/2*\mathrm{Z}/3$.

One important ingredient for the theorem of Johnson and Rees is the Kurosh subgroup theorem, according to which $\Gamma 1 * \Gamma 2$ would have a finite index subgroup of the form $\mathrm{Z}*K$, which would therefore also be the fundamental group of a projective variety. Another proof is based on the construction of a flat bundle Ecorresponding to a homomorphism $\mathrm{Z}*K \to \{\pm 1\}$ with some cohomology group $Hi(X,E)$ of odd dimension, contradicting the extension of Hodge theory to flat rank 1 bundles (see [181]).

Arapura, Bressler and Ramachandran answer in particular one question raised by Johnson and Rees, namely they show [7]:

Theorem 103 (Arapura, Bressler and Ramachandran) If X is a compact Kähler manifold, then its universal cover $\tilde{X}$ has only one end. And $\pi 1(X)$ cannot be a free product amalgamated by a finite subgroup.

Here, the ends of a non compact topological space X' are just the limit, as the compact subset K gets larger, of the connected components of the complement set $X' \backslash K$.

The theorem of Johnson and Rees admits the following consequence

Corollary 104 (Johnson–Rees) Let $X1$,$X2$ be smooth projective manifolds of dimension $n \geq 2$ with non zero first Betti number, or more generally with the property that $\pi 1(Xj)$ has a non trivial finite quotient. Then $X1 \sharp X2$ cannot be homeomorphic to a projective manifold.

It is interesting to observe that the previous result was extended by Donaldson [138] also to the case where $X1$,$X2$ are simply connected smooth projective surfaces. One could believe that the combination of the two results implies tout court that the connected sum $X1 \sharp X2$ of two projective manifolds of dimension at least two cannot be homeomorphic to a projective manifold. This is unfortunately not clear (at least to the author), because of a deep result by Toledo [354], giving a negative answer to another question posed by Serre

Theorem 105 (Toledo) There exist projective manifolds X whose fundamental group $G := \pi 1(X)$ is not residually finite.

This means that the natural homomorphism of $G := \pi 1(X)$ into its profinite completion $\hat{}G$ is not injective $\widehat{\pi_1(X)}$ is also called the algebraic fundamental group and denoted by $\pi 1(X)alg$.

The profinite completion is defined as the following inverse limit

$$\hat{G} := \lim_{K < G,} G/K,$$

where K runs through the set of normal subgroups of finite index.

Even the following question does not have yet a positive answer.

Question 106 Does there exist a projective manifold X such that $G = \pi 1(X) \neq 0$, while $\hat{}G =: \pi 1(X)alg = 0$? ($\pi 1(X)$ would then be an infinite group).

If the answer were negative, then the hypothesis in Corollary 104 would be that $X1$,$X2$ are not simply connected.

Concluding this section, it is clear how the results of Siu (and Mostow [300]), can be used to obtain restrictions for Kähler groups.

For instance, using the above mentioned techniques of classifying spaces and harmonic maps, plus some Lie theoretic arguments, Carlson and Toledo [75], while giving alternative proofs of the results of Siu, prove:

Theorem 107 (Carlson–Toledo) If X is a compact Kähler manifold, then its fundamental group $\pi 1(X)$ cannot be isomorphic to a discrete subgroup $\Gamma < SO(1,n)$ such that D/Γ be compact, where D is the hyperbolic space $SO(1, n) / SO(n)$.

Similar in spirit is the following theorem of Simpson [341], for which we recall that a lattice Γ contained in a Lie group G is a discrete subgroup such that G/Γ has finite volume, and that a reductive Lie group is of Hodge type if it has a compact Cartan subgroup.

Theorem 108 (Simpson) If X is a compact Kähler manifold, its fundamental group $\pi 1(X)$ cannot be isomorphic to a lattice Γcontained in a simple Lie group not of Hodge type.

For more information on Kähler groups (fundamental group of a cKM X), we refer to the book [2], to the survey article by Campana [72] and to [73].

7.6 Kähler versus projective, Kodaira's problem and Voisin's negative answer

Many of the restrictions for a group Γ to be a projective group (fundamental group of a projective variety X) are indeed restrictions to be a Kähler group. For long time, it was not clear which was the topological difference between projective smooth varieties and compact Kähler manifolds. Even more, there was a question by Kodaira whether any compact Kähler manifold X would be a deformation (or even a direct deformation) of a projective manifold Y, according to the following well known definition.

Definition 109 *(1)* Given two compact complex manifolds Y, X, Y is a **direct deformation** of X if there is a smooth proper and connected family $p{:}\mathcal{X}\to B$, where B is a smooth connected complex curve, such that X and Yisomorphic to some fibres of p.

*(2)*We say instead that Y is **deformation equivalent** to X if Y is equivalent to X for the equivalence relation generated by the (symmetric) relation of direct deformation.

Indeed, Kodaira proved:

Theorem 110 (Kodaira [249]) Every compact Kähler surface is a deformation of a projective manifold.

It was for a long time suspected that Kodaira's question, although true in many important cases (tori, surfaces), would be in general false. The counterexamples by Claire Voisin were based on topological ideas, namely on the integrality of some multilinear algebra structures on the cohomology of projective varieties.

To explain this, recall that a Kähler form ξ on a Kähler manifold X determines the Lefschetz operator on forms of type (p, q):

$$L : \mathcal{A}^{p,q} \to \mathcal{A}^{p+1,q+1}, \quad L(\phi) := \xi \wedge \phi$$

whose adjoint

$$\Lambda := L^* : \mathcal{A}^{p,q} \to \mathcal{A}^{p-1,q-1}$$

is given by $\Lambda = w * L *$ where

$$w := \sum_{p,q} (-1)^{p-q} \pi_{p,q}, \quad (Weil),$$
$$c := \sum_{p,q} (i)^{p-q} \pi_{p,q}, \quad (Chern - Weil),$$

are the Weil and Chern–Weil operators respectively. Here $\pi p,q$ is the projector onto the space of forms of type (p, q).

These operators satisfy the commutation relations:

$$(1)\ [L, w] = [L, c] = [\Lambda, w] = [\Lambda, c] = 0$$
$$(2)\ \ B := [L, \Lambda] = \sum (p + q - n) \pi_{p,q},$$
$$(3)\ [L, B] = -2L.$$

In this way the bigraded algebra of differential forms is a representation of the Lie Algebra $sl(2,\mathbb{C})$, which has the basis:

$$b = \begin{pmatrix} -1 & 0 \\ 0 & 1 \end{pmatrix},$$
$$l = \begin{pmatrix} 0 & 0 \\ 1 & 0 \end{pmatrix},$$
$$\lambda = \begin{pmatrix} 0 & 1 \\ 0 & 0 \end{pmatrix},$$

and with Lie bracket given by

$$b = [l, \lambda],$$
$$-2l = [l, b],$$
$$2\lambda = [\lambda, b].$$

For finite dimensional representations of $sl(2,C)$ one has the following structural result.

Proposition 111 Let $\rho{:}sl(2,C)\longrightarrow End(W)$ be a representation of $sl(2,C)$, $dim(W)<\infty$, and set $L{:=}\rho(l)$, $\Lambda{:=}\rho(\lambda)$, $B{:=}\rho(b)$. Then we have:

(1)$W=\bigoplus_{v\in Z}W_v$ is a finite direct sum, where W_v is the eigenspace of B with eigenvalue v.

(2)$L(W_v)\subset W_{v+2}$, $\Lambda(W_v)\subset W_{v-2}$.

(3)Let $P=\{w \mid \Lambda w=0\}$, be the space of primitive vectors, then we have a direct sum decomposition

$$W = \oplus_{r\in \mathbb{N}} L^r(P).$$

(4)Moreover, the irreducible representations of $sl(2,C)$ are isomorphic to $S^m(C^2)$, where $S^m(C^2)$ is the m-th symmetric power of the natural representation of $SL(2,C)$ on C^2 (i.e., the space of homogeneous polynomials of degree m).

(5)Let $P_\mu{:=}P\cap W_\mu$: then $P_\mu=0$ for $\mu>0$ and

$$P = \oplus_{\mu\in\mathbb{Z},\ \mu\le 0} P_\mu.$$

(6)$L^r{:}P_{-m}\longrightarrow W_{-m+2r}$ is injective for $r\le m$ and zero for $r>m$.

(7)$L^r{:}W_{-r}\longrightarrow W_r$ is an isomorphism $\forall r$, and

$$W_\mu = \oplus_{r\in \mathbb{N},\ r\ge\mu} L^r(P_{\mu-2r}).$$

On a compact Kähler manifold the projectors $\pi p,q$ commute with the Laplace operator, so that the components $\psi p,q$ of a harmonic form ψ are again harmonic, hence the operators $L,\Lambda,...$ preserve the finite dimensional subspace H of harmonic forms, and one can apply the above Proposition 111. The following theorem goes often after the name of Lefschetz decomposition, and is essential in order to prove the third Lefschetz theorem (see 3).

Theorem 112 *(i)* Let (X,ξ) be a compact Kähler manifold. Then Δ commutes with the operators $*, \partial, \bar{\partial}, \partial*, \bar{\partial}*, L, \Lambda$.

(ii)In particular, a k- form $\eta = \sum_{p+q=k} \eta_{p,q}$ is harmonic if and only if all the $\eta p,q$'s are harmonic.

(iii)Hence there is the so-called **Hodge Decomposition** of the de Rham cohomology:

$$\oplus_k H^k_{DR}(X, \mathbb{C}) \cong \oplus_k \mathcal{H}^k(X, \mathbb{C}) = \oplus_{p,q} \mathcal{H}^{p,q},$$

where Hp,q is the space of the harmonic Forms of Type (p, q),

(iv)**Hodge Symmetry** holds true, i.e., we have $\mathcal{H}^{p,q} = \overline{\mathcal{H}^{q,p}}$.

(v)There is a canonical isomorphism $\mathcal{H}^{p,q} \cong H^q(X, \Omega^p_X)$.

(vi)The Lefschetz operators L,Λ make H a representation of sl(2,C) 9, with eigenspaces $H\mu{:=}\bigoplus p{+}q{=}\mu{+}nHp,q$, and the space Pp,q of primitive forms of type (p, q) allows a decomposition

$$\mathcal{H}^{p,q} = \oplus_{r\in\mathbb{N},\ r\geq\mu=p+q-n} L^r(P_{p-r,q-r}),$$

where Pp,q=0 for $p+q>n$.

Moreover, the Hermitian product on Pp,q given by

$\langle\eta,\eta\rangle{:=}in-2pqL(n-p-q)(\eta\wedge\bar{}\eta)$

is positive definite.

Now, a projective manifold is a compact Kähler manifold when endowed with the Fubini-Study metric

$$-\frac{1}{2\pi i}\partial\bar{\partial}log\left(\sum |z_i|^2\right).$$

This is the Chern class $c1(L)$ where L is the line bundle OX(1).

Kodaira's embedding theorem characterizes projective manifolds as those compact Kähler manifolds which admit a Kähler metric whose associated form is integral, i.e., it is the first Chern class of a positive line bundle.

Definition 113 Let M be a compact differentiable manifold of real even dimension dimR(M)=2n.

(1)M is said to admit a real polarized Hodge structure if there exists a real closed two form ξ and a Hodge decomposition of its cohomology algebra (i.e., as in iii of Theorem 112) such that the operators $L(\psi):=\xi\wedge\psi$, and $\Lambda:w*L*$ satisfy properties (iv), (v) and (vi) of Theorem 112.

(2)M is said to admit an integral polarized Hodge structure if one can take ξ as above in $H2(M$,Z).

Thus the differentiable manifolds underlying a complex Kähler manifolds admit a real polarized Hodge structure, while those underlying a complex projective manifold admit an integral polarized Hodge structure.

Voisin showed that these properties can distinguish between Kähler and projective manifolds. We need here perhaps to recall once more that deformation equivalent complex manifolds are diffeomorphic.

Theorem 114 (Voisin [364]) In any complex dimension $n\geq 4$ there exist compact Kähler manifolds which are not homotopically equivalent to a complex projective manifold.

The construction is not so complicated: taking a complex torus T with an appropriate endomorphism φ, and blowing up $T\times T$ along four subtori

$$T \times \{0\}, \{0\} \times T, \Delta_T, \Gamma_\varphi,$$

where ΔT is the diagonal, and $\Gamma\varphi$ is the graph of φ, one obtains the desired Kähler manifold X.

The crucial property is the following lemma

Lemma 115 (Voisin [364]) Assume that the characteristic polynomial f of φ, a monic integral polynomial, has all eigenvalues of multiplicity 1, none of them is real, and moreover the Galois group of its splittiing field is the symmetric group S2*n*.

Then T is not an Abelian variety.

We refer to the original paper for details of the construction and proof. However, since our X is obviously bimeromorphic to the torus $T \times T$, which is a deformation of a projective manifold, later Voisin went on to prove a stronger result in [365].

Theorem 116 (Voisin [365]) In any complex dimension $n \geq 10$ there exist compact Kähler manifolds such that no compact smooth bimeromorphic model X' of X is homotopically equivalent to a complex projective manifold. Indeed, no such model X' has the same rational cohomology ring of a projective manifold.

That the ℝ rational cohomology ring was the essential topological invariant of Kähler manifolds, had been found already some 30 years before by Deligne, Griffiths, Morgan, and Sullivan [134], using Sullivan's theory of rational homotopy theory and minimal models of graded differential algebras (see [195] for a thourough introduction, and for some of the notation that we shall introduce here without explanation).

Theorem 117 Let X, Y be compact complex manifolds, which are either Kähler, or such that the *ddc* lemma holds for them (as for instance it happens when X is bimeromorphic to a cKM).

(1)Then the real homotopy type of X is determined by its real cohomology ring $H*(X,\mathrm{R})$, and similarly the effect of every holomorphic map $f{:}X \to Y$ on real homotopy types is a formal consequence of the induced homomorphism of real cohomology rings $f*{:}H*(Y,\mathrm{R}) \to H*(X,\mathrm{R})$.

(2)If moreover X is simply connected, then the graded Lie algebra of real homotopy groups $\pi*(X) \otimes$ZRdepends only on the cohomology ring $H*(X,\mathrm{R})$. In particular, all Massey products of any order are zero over Q.

(3)If instead X is not simply connected, then the real form of the canonical tower of nilpotent quotients of $\pi 1(X)$ (the real form is obtained by taking $\otimes$ZR of graded pieces and extension maps) is completely determined by $H1(X,\mathrm{R})$ and the cup product map

$$H^1(X, \mathbb{R}) \times H^1(X, \mathbb{R}) \to H^2(X, \mathbb{R}).$$

The article [134] contains several proofs, according to the taste of the several authors.

A basic property of compact Kähler manifolds which plays here a key role is the following lemma, also calledprinciple of the two types in [194].

Lemma 118

(ddc-Lemma) Let ϕ be a differential form such that

(1) $\partial\phi = \bar{\partial}\phi = 0$ (equivalently, $d\phi=dc\phi=0$, where dc is the real operator $-i(\partial - \bar{\partial})$),

(2)$d(\phi)=0$ (or $dc(\phi)=0$), then

(3)there exists ψ such that $\phi=\partial\bar{\partial}\psi$ (equivalently, there exists ψ such that $\phi=ddc(\psi)$).

One can see the relation of the above lemma with the vanishing of Massey products, which we now define in the simplest case.

Definition 119 Let X be a topological space, and let $a,b,c \in H^i(X,R)$ be cohomology classes such that

$$a \cup b = b \cup c = 0,$$

where R is any ring of coefficients.

Then the triple Massey product $\langle a,b,c\rangle$ is defined as follows: take cocycle representatives a',b',c' for$a$, b, c, respectively, and cochains $x,y \in C^{2i-1}(X,R)$ such that

$$dx = a' \cup b', dy = b' \cup c'.$$

Then $\langle a,b,c\rangle$ is the class of

$$a' \cup y + (-1)^{i+1} x \cup c'$$

inside

$$H^{3i-1}(X,R)/(a \cup H^{2i-1}(X,R) + H^{2i-1}(X,R) \cup c).$$

Concerning the choice of the ring R of coefficients: if we take X a cKM, and R=R, or R=Q, then these Massey products vanish, as stated in [134].

Instead things are different for torsion coefficients, as shown by Torsten Ekedahl [150].

Theorem 120 *(i)* There exist a smooth complex projective surface X and $a,b,c \in H1(X,Z/l)$ such that $a \cup b = b \cup c = 0$, but $\langle a,b,c \rangle \neq 0$.

(ii)Let $X \to P6$ be an embedding and let Y be the blow up of P6 along X.

Then there exist $a'',b'',c'' \in H3(X,Z/l)$ such that $a'' \cup b'' = b'' \cup c'' = 0$, but $\langle a'',b'',c'' \rangle \neq 0$.

Remark 121 *(1)*As already mentioned, Theorem 112 is an essential ingredient in the proof of the third Lefschetz theorem. Indeed, for a smooth hyperplane section $W = H \cap X$ of a projective variety X, the cup product with the hyperplane class $h \in H2(X,Z)$ corresponds at the level of forms to the operator L given by wedge product with ξ, the first Chern form of O$X(H)$.

(2)Most Kähler manifolds, for instance a general complex torus, do not admit any nontrivial analytic subvariety (nontrivial means: different from X or from a point). Whereas for projective varieties X of dimension $n \geq 2$ one can take hyperplane sections successively and obtain a surface S with $\pi 1(S) \cong \pi 1(X)$. Hence the well known fact that the set of fundamental groups of smooth projective varieties is just the set of fundamental groups of smooth algebraic surfaces.

(3)As explained in the book by Shafarevich [337], the same idea was used by J.P. Serre to show that any finite group G occurs as the fundamental group of a smooth projective surface S. Serre considers the regular representation on the Cartesian product of Y indexed by G, which we denote as usual by YG, and where for instance Y=Pm is a simply connected projective variety of dimension $m \geq 3$. The quotient $X := (YG)/G$ is singular on the image of the big diagonal, which however has codimension at least m. Cutting$X$ with an appropriate number of general hyperplanes one obtains a surface S with $\pi 1(S) \cong G$.

(4)Serre's result is used in Ekedahl's theorem: Ekedahl shows the existence of a group G of order $l5$ such that there is a principal fibration $K(G,1) \to K((Z/l)3,1)$ with fibre a $K((Z/l)2,1)$; moreover he deduces the existence of elements a, b, c as desired from the fact that there is some CW complex with non zero triple Massey product.

Then the blow up of P6 along a surface X with $\pi 1(X) \cong G$ is used to obtain an example with classes in the third cohomology group $H3(Y,Z/l)$.

7.7 The Shafarevich conjecture

One of the many reasons of the beauty of the theory of curves is given by the uniformization theorem, stating that any complex manifold C of dimension 1 which is not of special type (i.e., not P1, C, C$*$, or an elliptic curve) has as universal covering the unit disk B1={z∈C||z|<1}, which is biholomorphic to the upper half plane H={z∈C|$Im(z)$>0}.

In other terms, the universal cover is either P1, C, or H according to the sign of the curvature of a metric with constant curvature (positive, zero or negative).

In higher dimensions there are simply connected projective varieties that have a positive dimensional moduli space: already in the case of surfaces, e.g., smooth surfaces in P3, of degree at least 3, there is an uncountable family of pairwise non isomorphic varieties.

So, if there is some analogy, it must be a weaker one, and the first possible direction is to relate somehow Kodaira dimension with curvature.

In the case where the canonical divisor KX is ample, there is the theorem of Aubin and Yau (see [16, 376]) showing the existence, on a projective manifold with ample canonical divisor KX, of a Kähler–Einstein metric, i.e. a Kähler metric ω such that

$$Ric(\omega)=-\omega.$$

This theorem is partly the right substitute for the uniformization theorem in dimension $n>1$, but the Kähler–Einstein condition, i.e. the existence of a Kähler metric ω such that

$$Ric(\omega)=c\omega,$$

forces KX ample if $c<0$, KX trivial if $c=0$, and $-KX$ ample if $c>0$.

In the case where KX is trivial, Yau showed the existence of a Kähler–Einstein metric in [376], while the existence of such a metric on Fano manifolds (those with ample anticanonical divisor $-KX$) has only recently been settled, under a stability assumption (see [122, 123, 352]).

In the general case, one is looking for suitable metrics on singular varieties, but we shall not dwell on this here, since we are focusing on topological aspects: we refer for instance to [162] (see also [204]). Once more, however, curvature influences topology: Yau showed in fact [375] that, for a projective manifold with ample canonical divisor *KX*, the famous Yau inequality is valid

$$K_X^n \leq \frac{2(n+1)}{n} K_X^{n-2} c_2(X),$$

equality holding if and only if the universal cover ˜*X* is the unit ball B*n* in C*n*.

The second possible direction is to investigate properties of the universal covering ˜*X* of a projective variety. For instance, one analogue of projective curves of genus *g*≥2, whose universal cover is the unit ball B1⊂C, is given by the compact complex manifolds *X* whose universal covering ˜*X* is biholomorphic to a bounded domain Ω⊂C*n*.

Necessarily such a manifold *X* is projective and has ample canonical divisor *KX* (see [246, 247], Theorem 8.4 p. 144, where the Bergman metric is used, while the method of Poincaré series is used in [340], Theorem 3 p. 117 , see also [255], Chapter 5).

Moreover, it is known, by a theorem of Siegel ([339], cf. also [242], Theorem 6.2), that Ω must be holomorphically convex, indeed a Stein domain. We recall these concepts (see for instance [6, 185]).

Definition 122 A complex space *X* is said to be a Stein complex space if and only if one of the following equivalent conditions hold:

- (1S) *X* is a closed analytic subspace of C*N*, for some *N*,
- (2S) for any coherent analytic sheaf F, we have *Hj*(*X*,F)=0, for all *j*≥1,
- (3S) *X* coincides with the maximal spectrum *Spec^m*(*HolX*) of its algebra of global holomorphic functions *HolX*:=*H*0(*X*,O*X*), where *Specm*(*HolX*) is the subspace of *Spec*(*HolX*) consisting of the maximal ideals.

- (4S) Global holomorphic functions separate points and X possesses a C^∞ strictly plurisubharmonic exhaustion function, i.e., there is a real valued proper function $f:X\to \mathbb{R}$ such that the Levi form $L(f):=i\bar{\partial}\partial f$ is strictly positive definite on the Zariski tangent space of each point.

A complex space is said to be holomorphically convex if and only if one of the following equivalent conditions hold:

- (1HC) For each compact K, its envelope of holomorphy $K':=\{x\in X \mid |f(x)|\le \max_K|f|\}$ is also compact

- (2HC) X admits a proper holomorphic map $s:X\to\Sigma$ to a complex Stein space (this map is called Steinification, or Cartan–Remmert reduction), which induces an isomorphism of function algebras $s^*:Hol\,\Sigma\to Hol\,X$.

- (3HC) X admits a real valued twice differentiable proper function $f:X\to\mathbb{R}$ and a compact $K\subset X$ such that

(3i) there is a number $c\in\mathbb{R}$ with $f^{-1}((-\infty,c])\subset K$

(3ii) the Levi form $L(f):=i\bar{\partial}\partial f$ is semipositive definite and strictly positive definite outside K.

Finally, a complex space X is Stein manifold iff it is holomorphically convex and global holomorphic functions separate points.

Now, a compact complex manifold (or space) is by definition holomorphically convex, so this notion captures two extreme behaviours of the universal covers we just described: bounded domains in $\mathbb{C}^n$, and compact manifolds. In his book [337] Shafarevich answered the following

Shafarevich's question: is the universal covering $\tilde{X}$ of a projective variety X, or of a compact Kähler manifold X, holomorphically convex?

Remark 123 The answer is negative for a compact complex manifold that is not Kähler, since for instance Hopf surfaces X=(C2\{0})/Z have C2\{0} as universal cover, which is not Stein since it has the same algebra of holomorphic functions as the larger affine space C2.

Observe that all the $K(\pi,1)$ projective manifolds we have considered so far satisfy the Shafarevich property, since their universal covering is either Cn or a bounded domain in Cn (this holds also for Kodaira fibred surfaces, by Bers' simultaneous uniformization, [40]).

An interesting question concerns projective varieties whose universal cover is a bounded domain D in Cn. In this case the group Aut(D) contains an infinite cocompact subgroup, so it is natural to look first at domains which have a big group of automorphisms, especially at **bounded homogeneous domains**, i.e., bounded domains such that the group Aut(Ω) of biholomorphisms of Ω acts transitively.

But, as already mentioned, a classical result of J. Hano (see [207] Theorem IV, p. 886, and Lemma 6.2, p. 317 of [292]) asserts that a bounded homogeneous domain that covers a compact complex manifold is a bounded symmetric domain.

These naturally occur as such universal covers: Borel proved in [57] that for each bounded symmetric domain there exists a compact free quotient X=Ω/Γ, called a Hermitian locally symmetric projective manifold (these were also called compact Clifford-Klein forms of the symmetric domain Ω).

By taking finite ramified coverings of such locally symmetric varieties, and blow ups of points of the latter, we easily obtain many examples which are holomorphically convex but not Stein. But if we blow up some subvariety Y⊂X of positive dimension some care has to be taken: what is the inverse image of Y in the universal cover ˜X? It depends on the image of π1(Y) inside π1(X): if the image is finite, then we obtain a disjoint union of compact varieties, else the connected components of the inverse image are not compact.

A more general discussion leads to the following definition.

Definition 124 Assume that the Shafarevich property holds for a compact Kähler manifold: then the fundamental group π1(X)acts properly discontinuously on ˜X and on its Steinification Σ, hence one has a quotient complex space of Σwhich is holomorphically dominated by X:

$$Shaf(X) := \Sigma/\pi_1(X), shaf_X : X = \tilde{X}/\pi_1(X) \to Shaf(X) = \Sigma/\pi_1(X).$$

Shaf(X) is then called the Shafarevich variety of X, and *shafX* is called the Shafarevich morphism. They are characterized by the following universal property:

a subvariety $Y \subset X$ is mapped to a point in *Shaf*(X) if and only if, letting Y^n be the normalization of Y, the image of $\pi_1(Y^n) \to \pi_1(X)$ is finite.

So, a first attempt towards the question was in the 90's to verify the existence of the Shafarevich morphism *shafX*.

A weaker result was shown by Kollár [254] and by Campana [71]: we borrow here the version of Kóllar, which we find more clearly formulated.

Theorem 125 Let X be a normal projective variety: then X admits a rational Shafarevich map $shaf_X : X \dashrightarrow Shaf(X)$, with the following properties:

(1)*shafX* has connected fibres

(2)there are countably many subvarieties $D_i \subset X$ such that for every subvariety $Y \subset X$, $Y \not\subset \cup_i D_i$, $shaf_X(Y)$ is a point if and only if, letting Y^n be the normalization of Y, the image of $\pi_1(Y^n) \to \pi_1(X)$ is finite.

Moreover, the above properties determine *Shaf*(X) up to birational equivalence.

Remark 126**(1)** A subtle point concerning the definition of Shafarevich morphism is the one of passing to the normalization of a subvariety Y. In fact, the normalization Y^n has a holomorphic map to X since X is normal, and the fibre product $Y^n \times_X \tilde{X}$ has a holomorphic map to $\tilde{X}$. If the image of $\pi_1(Y^n) \to \pi_1(X)$ is finite, this fibre product consists of a (possibly infinite) Galois unramified covering of Y^n whose components are compact.

If $\tilde{X}$ is holomorphically convex, then these components map to points in the Steinification Σ.

(2) Consider the following special situation: X is a smooth rational surface, and Y is a curve of geometric genus zero with a node. Then the normalization is $\mathbb{P}^1$ and the Shafarevich map should contract Y^n.

Assume however that $\pi_1(Y) \cong \mathbb{Z}$ injects into $\pi_1(X)$: then the inverse image of Y in $\tilde{X}$ consists of an infinite chain of $\mathbb{P}^1$'s, which should be contracted by the Steinification morphism, hence this map would not be proper. The existence of a curve satisfying these two properties would then give a negative answer to the Shafarevich question.

(3) In other words, the Shafarevich property implies that:

(***) if $\pi 1(Yn) \to \pi 1(X)$ has finite image , then also $\pi 1(Y) \to \pi 1(X)$ has finite image.

So, we can define a subvariety to be **Shafarevich bad** if

(SB) $\pi 1(Yn) \to \pi 1(X)$ has finite image , but $\pi 1(Y) \to \pi 1(X)$ has infinite image.

The rough philosophy of the existence of a rational Shafarevich map is thus that Shafarevich bad subvarieties do not move in families which fill the whole X, but they are contained in a countable union of subvarieties Di(one has indeed to take into account also finite union of subvarieties, for instance two P1's crossing transversally in 2 points: these are called by Kollár 'normal cycles').

But still, could they exist, giving a negative answer to the Shafarevich question?

Potential counterexampes were proposed by Bogomolov and Katzarkov [49], considering fibred surfaces $f{:}X \to B$. Indeed, for each projective surface, after blowing up a finite number of points, we can always obtain such a fibration.

Assume that $f{:}X \to B$ is a fibration of hyperbolic type: then, passing to a finite unramified covering of X and passing to a finite covering of the base B, we may assume that the fibration does not have multiple fibres.

Then (see Theorem 96) the surjection $\pi 1(X) \to \pi 1(B)$ induces an infinite unramified covering

$\hat{}f{:}\hat{}X \to \hat{}B$:=H=B1,

and we shall say that this fibration is obtained by *opening the base.*

If instead $f{:}X \to B$ is a fibration of parabolic type, we get an infinite unramified covering

$\hat{}f{:}\hat{}X \to \hat{}B$:=C.

In the elliptic case, B=P1 and we set $\hat{}X{:=}X, \hat{}B{:=}B$.

Now, the Shafarevich question has a positive answer if the fundamental group of ˆX, the image of the fundamental group of a general fibre F inside $\pi1(X)$, is finite.

Assume instead that this image is infinite and look at components of the singular fibres: we are interested to see whether we find some Shafarevich bad cycles. For this, it is important to describe the fundamental group $\pi1(\hat{}X)$ as a quotient of $\pi g=\pi1(F)$, where F is a general fibre, and g is its genus.

Observe preliminarily that we may assume that there are singular fibres, else either we have a Kodaira fibration, or an isotrivial fibration (see [285]) so that, denoting by $B*$ the complement of the critical locus of ˆf, the fundamental group of $B*$ is a free group.

For each singular fibre $Ft,t\in B$, a neighbourhood of Ft is retractible to Ft, and because of this we have a continuous map $F\rightarrow Ft$.

The **group of local vanishing cycles** is defined as the kernel of $\pi1(F)\rightarrow\pi1(Ft)$,and denoted by $Vant$.

Then we obtain a description of $\pi1(\hat{}X)$ as

$$\pi_1(\hat{X}) = \pi_g/\langle\langle \cup_t \{Van_t\}\rangle\rangle,$$

where $\langle\langle M\rangle\rangle$ denotes the subgroup normally generated by M.

Bogomolov and Katzarkov consider the situation where the fibre singularities are exactly nodes, and then for each node p there is a vanishing cycle $vanp$ on the nearby smooth fibre. In this case we divide by the subgroup normally generated by the vanishing cycles $vanp$.

Their first trick is now to replace the original fibration by the pull back via a map $\varphi:B'\rightarrow B$ which is ramified at each critical value t of multiplicity N.

In this way they obtain a surface X' which is singular, with singular points q of type $AN-1$, i.e., with local equation $zN=xy$. Since these are quotient singularities C2/(Z/N), they have local fundamental group $\pi1,loc(X'\setminus\{q\})$=Z/$N$.

As a second step, they construct another surface SN such that the image of $\pi1(SN)$ has finite index inside $\pi1(X'\setminus Sing(X'))$.

They show (lemma 2.7 and theorem 2.3 of loc. cit.) the following.

Proposition 127 The Bogomolov–Katzarkov procedure constructs a new fibred surface *SN* such that, if we open the base of the fibration, we get a surface ^*SN* such that

$$\pi_1(\hat{S_N}) = \pi_g/\langle\langle \cup_p \{van_p^N\}\rangle\rangle.$$

In this way they propose to construct counterexamples to the Shafarevich question (with non residually finite fundamental group), provide certain group theoretic questions have an affirmative answer.

Remark 128 We want to point out a topological consequence of the Shafarevich property, for simplicity we consider only the case of a projective surface *X* with infinite fundamental group.

Assume that ˜*X* is holomorphically convex, and let *s*:˜*X*→Σ the Cartan–Remmert reduction morphism, where Σ is Stein and simply connected. If *dim*(Σ)=1, then Σ is contractible, otherwise we know [4] that Σ is homotopy equivalent to a CW complex of real dimension 2.

In the first case we have a fibration with compact fibres of complex dimension 1, in the second case we have a discrete set such that the fibre has complex dimension 1, and the conclusion is that:

if a projective surface with infinite fundamental group satisfies the Shafarevich property, then ˜*X* is homotopy equivalent to a CW complex of real dimension at most 2.

To my knowledge, even this topological corollary of the Shafarevich property is yet unproven.

On the other hand, the main assertion of the Shafarevich property is that the quotient of ˜*X* by the equivalence relation which contracts to points the compact analytic subspaces of ˜*X* is not only a complex space, but a Stein space. This means that one has to produce a lot of holomorphic functions on ˜*X*, in order to embed the quotient as a closed analytic subspace in some C*N*.

There are positive results, which answer the Shafarevich question in affirmative provided the fundamental group of *X* satisfies certain properties, related somehow to some of the themes treated in this article, which is the existence of certain homomorphisms to fundamental groups of classifying spaces, and to the theory of harmonic maps. To give a very very simple idea, an easy result in this directions is that ˜*X* is a Stein manifold if the Albanese morphism *α*:*X*→*A*:=*Alb*(*X*) is a finite covering.

Now, since there has been a series of results in this direction,we refer to the introduction and bibliography in the most recent one [160] for some history of the problem and for more information concerning previous results, especially the first ones due to Jost and Zuo (for instance, [198, 229–231]) and other ones.

We want however to directly cite some previous results due to Katzarkov and Ramachandran [234], respectively to Eyssidieux [159] (see also [161] for a general introduction) which are simpler to state. We only want to recall that a reductive representation is one whose image group is reductive, i.e., all of its representations are semi-simple, or completely reducible (any invariant subspace $W \subset V$has an invariant complement).

Theorem 129 (Katzarkov–Ramachandran) Let X be a normal Kähler compact surface, and $X' \to X$ an unramified cover with Galois group Γ (so $X = X'/\Gamma$). If Γ does not contain Z as a finite index subgroup and it admits an almost faithful (i.e., with finite kernel) Zariski dense representation in a connected reductive complex Lie group, then X' is holomorphically convex.

Theorem 130 (Eyssidieux) Let X be a smooth projective variety, consider a homomorphism $\rho : \pi 1(X) \to G$, and let $\tilde{}X\rho := \tilde{}X/ker(\rho)$ be the connected unramified covering of X associated to $ker(\rho)$.

(1)If G=GL(n,C) and ρ is a reductive representation, then there is a relative Shafarevich morphism

$shaf\rho : X \to Shaf\rho(X)$

to a normal projective variety, satisfying the universal property that for each normal variety Z mapping to X, its image in $Shaf\rho(X)$ is a point if and only if the image of $\pi 1(Z) \to \pi 1(X)$ is finite. Moreover, the connected components of the fibres of $\tilde{}X\rho \to Shaf\rho(X)$ are compact.

(2)Let M be a quasi compact and absolute constructible (i.e., it remains constructible after the action of each $\sigma \in Aut$(C)) set of conjugacy classes of reductive linear representations of $\pi 1(X)$, and let HM be the intersection of the respective kernels. Then $\tilde{}XM := \tilde{}X/HM$ is holomorphically convex.

The main result of [160] consists of two parts, the first one being the following:

Theorem 131 (Eyssidieux–Katzarkov–Pantev–Ramachandran) Let X be a smooth projective variety, and let G be a reductive algebraic group defined over Q. Consider the Betti character scheme $M:=MB(X,B)$ such that, for a C-algebra A of finite type, the set of A-valued points $M(A)=M(Spec(A))$ parametrizes the representations

(∗) $\rho:\pi 1(X)\rightarrow G(A)$.

(1)Denote by $\tilde{H}_M^{\infty}$ the kernel of all such representations: then the associated Galois connected unramified covering $\tilde{X}_M^{\infty}$ is holomorphically convex.

(2)there is a non-increasing family of normal subgroups ˜*HkM* (obviously containing ˜*H∞M*), which correspond to homomorphisms $\rho:\pi 1(X)\rightarrow G(A)$, with the property that A is an Artinian local C-algebra, and that the Zariski closure of the image has k-step unipotent radical.

Then the associated Galois connected unramified covering ˜*XkM* is holomorphically convex.

7.8 Strong and weak rigidity for projective $K(\pi,1)$ manifolds

We want here to see how the moduli problem is solved for many projective $K(\pi,1)$ manifolds, which were mentioned in Sect. 3. However, we want here not to be entangled in the technical discussion whether a variety of moduli exists, and, if so, which properties does it have. So, we use the notion of deformation equivalence introduced by Kodaira and Spencer.

Definition 132 (*Rigidity*)

(1)Let X be a projective manifold: then we say that X is **strongly rigid** if, for each projective manifold Yhomotopy equivalent to X, then Y is isomorphic to X or to the complex conjugate variety $\bar{X}$ ($X\cong Y$ or $\bar{X}\cong Y$).

(2)We say instead that X is **weakly rigid** if, for each other projective manifold Y homotopy equivalent to$X$, then either Y is **direct deformation** of X, or Y is **direct deformation** of $\bar{X}$ (see 109).

(3)We say instead that X is **quasi rigid** if, for each other projective manifold Y homotopy equivalent to X, then either Y is **deformation**

equivalent to X, or Y is **deformation equivalent** to $\bar{X}$ (recall that deformation equivalence is the equivalence relation generated by the relation of direct deformation).

Remark 133(a) This is the intuitive meaning of the above definition: assume that there is a moduli space M, whose points correspond to isomorphism classes of certain varieties, and such that for each flat family $p:\mathcal{X}\to B$ the natural map $B\to$M associating to $b\in B$ the isomorphism class of the fibre $X_b:=p^{-1}(b)$ is holomorphic. Then to be a direct deformation of each other means to belong to the same irreducible component of the moduli space M, while being deformation equivalent means to belong to the same connected component of the moduli space M.

(b)The same definition can be given in the category of compact Kähler manifolds, or in the category of compact complex manifolds.

(c)Of course, if $p:\mathcal{X}\to B$ is a smooth proper family with base B a smooth connected complex manifold, then all the fibres X_b are direct deformation equivalent.

(d)We shall need in the sequel some more technical generalization of these notions, which amount, given a product of varieties $X_1\times X_2\times\cdots\times X_h$, to take the complex conjugate of a certain number of factors.

Definition 134 Let X be a projective manifold: then we say that X is **strongly * rigid** if, for each other projective manifold Yhomotopy equivalent to X, then Y is isomorphic to X or to another projective variety Z obtained by X as follows: there is a Galois unramified covering X' of X (thus $X=X'/G$) which splits as a product

$$X' = X'_1 \times X'_2 \times \cdots \times X'_h.$$

Set then, after having fixed a subset $\mathcal{J} \subset \{1,\dots,h\}$: $Z'_j := \overline{X'_j}$ for $j\in$J, and $Z'_j:=X'_j$ for $j\notin$J. Consider then the action of G on $Z' = (Z'_1 \times Z'_2 \times \cdots \times Z'_h)$, and let $Z:=Z'/G$.

Replacing the word: ‘isomorphic’ by ‘direct deformation equivalent’, resp. ‘deformation equivalent’, we define the notion of **weakly * rigid**, resp. **quasi * rigid**.

(1)The first example, the one of projective curves, was essentially already discussed: we have the universal family

$$p_g : \mathcal{C}_g \rightarrow \mathcal{T}_g,$$

hence, according to the above definition, **projective curves are weakly rigid** (and in a stronger way, since we do not need to allow for complex conjugation, as the complex conjugate ¯*C* is a direct deformation of *C*).

(2)**Complex tori are weakly rigid in the category of compact Kähler manifolds**, since any manifold with the same integral cohomology of a complex torus is a complex torus (Theorem 77).

Moreover, complex tori are parametrized by an open set T*n* of the complex Grassmann Manifold *Gr*(*n*, 2*n*), image of the open set of matrices {Ω∈*Mat*(2*n*,*n*;C) |*indet*(Ω¯Ω)>0}, as follows: we consider a fixed lattice Γ≅Z2*n*, and to each matrix Ω as above we associate the subspace

$$V = \Omega\mathbb{C}^n,$$

so that *V*∈*Gr*(*n*,2*n*) and Γ⊗C≅*V*⊕¯*V*.

Finally, to Ω we associate the torus *YV*:=*V*/*pV*(Γ), *pV*:*V*⊕¯*V*→*V* being the projection onto the first addendum.

As it was shown in [100] (cf. also [103]) T*n* is a connected component of Teichmüller space.

(3)Complex tori are not quasi rigid in the category of compact complex manifolds. Sommese generalized some construction by Blanchard and Calabi, obtaining [348] that the space of complex structures on a six dimensional real torus is not connected.

(4)**Abelian varieties are quasi rigid**, but not weakly rigid. In fact, we saw that all the Abelian varieties of dimension *g* admitting a polarization of type *D*′ are contained in a family over H*g*. Moreover, products *E*1×⋯×*Eg* of elliptic curves admit polarizations of each possible type.

(5)Locally symmetric manifolds D/Γ where D is irreducible of dimension >1 are **strongly rigid** by Siu's theorem 90.

(6)Locally symmetric manifolds D/Γ with the property that D does not have any irreducible factor of dimension 1 are strongly * rigid by Siu's theorem 90.

(7)Varieties isogenous to a product (VIP) are **weakly * rigid** in all dimensions, according to Theorem 136, that we are going to state soon (see below). They are **weakly rigid** in dimension n=2 only if we require the homotopy equivalence to be orientation preserving.

(8)Among the VIP's, the **strongly * rigid** are exactly the quotients X=($C1\times C2\times\cdots\times Cn)/G$ where G not only acts freely, but satisfies the following property. Denote by $G0\subset G$ (see [96] for more details) the subgroup which does not permute the factors, and observe that

$$G^0 \subset \mathrm{Aut}(C_1) \times \mathrm{Aut}(C_2) \times \cdots \times \mathrm{Aut}(C_n).$$

Then $G0$ operates on each curve Ci, and the required condition is that this action is rigid, more precisely we want : $Ci/G0\cong$P1 and the quotient map pi:$Ci\to Ci/G0\cong$P1 is branched in three points (we shall also say that we have then a **triangle curve**).

Again, in dimension n=2 we get some strongly rigid surfaces, which have been called (ibidem) **Beauville surfaces**. We propose therefore to call **Beauville varieties** the strongly * rigid VIP's.

(9)Hyperelliptic surfaces are weakly rigid, as classically known, see below for a more general result.

(10)It is unclear, as we shall see, whether Kodaira fibred surfaces are weakly * rigid, however any surface homotopically equivalent to a Kodaira fibred surface is also a Kodaira fibred surface. Indeed a stronger result, with a similar method to the one of [96] was shown by Kotschick (see [258] and Theorem 138 below), after some partial results by Jost and Yau.

(11)Generalized hyperelliptic varieties X are a class of varieties for which a weaker property holds. Namely, if Y is a compact Kähler manifold which is homotopy equivalent to X (one can relax this assumption a bit, obtaining stronger results), then Y is the quotient of Cn by an affine action of $\Gamma:=\pi 1(X)\cong\pi 1(Y)$ which, using Proposition 150, can be shown to have the same real affine type as the action yielding X as a quotient. But the Hodge type could be different, except in special cases where weak rigidity holds.

The following (see [96, 98]) is the main result concerning surfaces isogenous to a product, and is a stronger result than weak * rigidity.

Theorem 135 (a) A projective smooth surface S is isogenous to a product of two curves of respective genera $g1,g2\geq 2$, if and only if the following two conditions are satisfied:

(1)there is an exact sequence

$$1 \to \pi_{g_1} \times \pi_{g_2} \to \pi = \pi_1(S) \to G \to 1,$$

where G is a finite group and where πgi denotes the fundamental group of a projective curve of genus $gi\geq 2$;

(2) $e(S)(= c_2(S)) = \frac{4}{|G|}(g_1 - 1)(g_2 - 1).$ Any surface X with the same topological Euler number and the same fundamental group as S is diffeomorphic to S and is also isogenous to a product. There is a smooth proper family with connected smooth base manifold T, p:X→T having two fibres respectively isomorphic to X, and Y, where Y is one of the 4 surfaces $S = (C_1 \times C_2)/G,\ S_{+-} := (\overline{C_1} \times C_2)/G.$, $\bar{S} = (\overline{C_1} \times \overline{C_2})/G,\ S_{-+} := (C_1 \times \overline{C_2})/G = \overline{S_{+-}}.$

(c) The corresponding subset of the moduli space of surfaces of general type $\mathfrak{M}_S^{top} = \mathfrak{M}_S^{diff}$, corresponding to surfaces orientedly homeomorphic, resp. orientedly diffeomorphic to S, is either irreducible and connected or it contains two connected components which are exchanged by complex conjugation.

In particular, if S' is orientedly diffeomorphic to S, then S' is deformation equivalent to S or to $\bar{S}$.

Idea of the proof $\Gamma:=\pi 1(S)$ admits a subgroup Γ' of index d such that $\Gamma'\cong(\pi g1\times\pi g2)$. Let S' be the associated unramified covering of S. Then application of the isotropic subspace theorem or of Theorem 96 yields a pair of holomorphic maps $fj:S'\to Cj$, hence a holomorphic map

$$F := f_1 \times f_2 : S' \to C_1' \times C_2'.$$

Then the fibres of $f1$ have genus $h2\geq g2$, hence by the Zeuthen Segre formula (2) $e(S')\geq 4(h2-1)(g1-1)$, equality holding if and only if all the fibres are smooth.

But $e(S')=4(g1-1)(g2-1)\leq 4(h2-1)(g1-1)$, so $h2=g2$, all the fibres are smooth hence isomorphic to $C'2$; therefore F is an isomorphism.

The second assertion follows from the refined Nielsen realization theorem, Theorem 37, considering the action of the index two subgroup

$G0$ of G. Notice that if a group, here $G0$, acts on a curve C, then it also acts on the complex conjugate curve, thus $G0\subset\mathrm{Aut}(C)\Rightarrow G0\subset\mathrm{Aut}(\bar{C})$. But since complex conjugation is orientation reversing, these two actions are conjugate by $\mathrm{Out}(\pi g)$, not necessarily by $\mathrm{M}apg=\mathrm{Out}^+(\pi g)$.

Recall now that the exact sequence

$1\to\pi g1\times\pi g2\to\Gamma0\to G0\to1$

yields two injective homomorphisms $\rho j{:}G0\to Out(\pi gj)$ for j=1,2. Now, we choose the isomorphism $\pi_1(C_j)\cong\pi_{g_j}$ in such a way that it is orientation preserving. The isomorphism $\pi1(X)\cong\pi1(S)$, and the realization $X=(C'1\times C'2)/G$ then gives an isomorphism $\phi j{:}\pi1(C'j)\cong\pi gj$. If this isomorphism is orientation preserving, then the two actions of $G0$ on $Cj,C'j$ are deformation of each other by the refined Nielsen realization. If instead this isomorphism is orientation reversing, then we replace Cj by $\bar{C}j$ and now the two actions are conjugate by $\mathrm{M}apg$, so we can apply the refined Nielsen realisation theorem to $\bar{C}j,C'j$. Finally, in the case where the homeomorphism of X with S is orientation preserving, then either both $\phi1,\phi2$ are orientation preserving, or they are both orientation reversing. Then X is a deformation either of S or of $\bar{S}$. ▯

The first part of the following theorem is instead a small improvement of theorem 7.1 of [96], the second part is theorem 7.7 ibidem, and relies on the results of Mok [294, 295].

Theorem 136 *(1)* Let Y be a projective variety of dimension n with KY ample and that $\Gamma{:=}\pi1(Y)$ admits a subgroup Γ' of index d such that

$$\Gamma'\cong\pi_{g_1}\times\pi_{g_2}\times\cdots\times\pi_{g_n},\ g_i\geq 2\ \ \forall i,$$

and moreover $H2n(\Gamma,\mathrm{Z})\to H2n(Y,\mathrm{Z})$ is an isomorphism. Then Y is a variety isogenous to a product.

(2)Let X be projective variety with universal covering the polydisk $\mathrm{H}n$, and let Y be a projective variety of dimension n with KY ample and such that $\pi1(Y)\cong\Gamma{:=}\pi1(X)$, and assume that $H^{2n}(\Gamma,\mathbb{Z})\to H^{2n}(Y,\mathbb{Z})$ is an isomorphism. Then also Y has $\mathrm{H}n$ as universal covering (hence we have a representation of Y as a quotient $Y=\mathrm{H}n/\Gamma$).

(3)In particular X is weakly * rigid, and strongly * rigid if the action of Γ is irreducible (this means that there is no finite index subgroup $\Gamma'<\Gamma$ and an isomorphism $\mathrm{H}n\cong\mathrm{H}m\times\mathrm{H}n-m$ such that

$$\Gamma' = \Gamma'_1 \times \Gamma'_2, \ \ \Gamma'_1 \subset Aut(\mathcal{H}^m), \ \ \Gamma'_2 \subset Aut(\mathcal{H}^{n-m}).$$

Proof of (1): As in the previous theorem we take the associated unramified covering Y' of Y associated to Γ', and with the same argument we produce a holomorphic map

$F{:}Y' \rightarrow Z{:=}C1{\times}C2{\times}\cdots{\times}Cn.$

We claim that F is an isomorphism. Indeed, by the assumption $H2n(\Gamma,\mathbb{Z}) \cong H^{2n}(Y,\mathbb{Z})$ follows that $H^{2n}(\Gamma',\mathbb{Z}) \cong H2n(Y',\mathbb{Z})$, hence F has degree 1 and is a birational morphism. Let R be the ramification divisor of F, so that $KY'{=}F{*}(KZ){+}R$. Now, since F is birational we have $H0(mKY') \cong H0(mKZ)$, for all $m \geq 0$. This implies for the corresponding linear systems of divisors:

$$|mK_{Y'}| = F^*(|mK_Z|) + mR.$$

For $m>>0$ this shows that KY' cannot be ample, contradicting the ampleness of KY.

Remark 137 Part (2) of the above theorem is shown in [96], while part (3) follows by the same argument given in the proof of Theorem 135 in the case where X is isogenous to a product. We omit the proof of (3) in general.

Theorem 138 (Kotschick) Assume that S is a compact Kähler surface, and that

(i)its fundamental group sits into an exact sequence, where $g,b \geq 2$:

$1 \rightarrow \pi g \rightarrow \pi 1(S) \rightarrow \pi b \rightarrow 1$

(ii)$e(S){=}4(b{-}1)(g{-}1)$.

Then S has a smooth holomorphic fibration $f{:}S \rightarrow B$, where B is a projective curve of genus b, and where all the fibres are smooth projective curves of genus g. f is a Kodaira fibration if and only if the associated homomorphism $\rho{:}\pi b \rightarrow \mathrm{M}apg$ has image of infinite order, else it is a surface isogenous to a product of unmixed type and where the action on the first curve is free.

Proof By Theorem 96 the above exact sequence yields a fibration f:S→B such that there is a surjection $\pi 1(F)$→πg, where F is a smooth fibre. Hence, denoting by h the genus of F, we conclude that h≥g, and again we can use the Zeuthen–Segre formula to conclude that h=g and that all fibres are smooth. So F is a smooth fibration. Let C'→C be the unramified covering associated to ker(ρ): then the pull back family S'→C' has a topological trivialization, hence is a pull back of the universal family Cg→Tg for an appropriate holomorphic map φ:C'→Tg.

If ker(ρ) has finite index, then C' is compact and, since Teichmüller space is a bounded domain in C3g−3, the holomorphic map is constant. Therefore S' is a product C'×$C$2 and, denoting by G:=Im(ρ), S=(C'×$C$2), and we get exactly the surfaces isogenous to a product such that the action of G on the curve C' is free.

If instead G:=Im(ρ) is infinite, then the map of C' into Teichmüller space is not constant, since the isotropy group of a point corresponding to a curve F is, as we saw, equal to the group of automorphisms of F (which is finite). Therefore, in this case, we have a Kodaira fibration. ▯

Remark 139 Jost and Yau [226] proved a weaker result, that a deformation of the original examples by Kodaira of Kodaira fibrations are again Kodaira fibrations. In our joint paper with Rollenske [109] we gave examples of Kodaira fibred surfaces which are rigid, and we also described the irreducible connected components of the moduli space corresponding to the subclass of those surfaces which admit two different Kodaira fibrations.

Question 140 Are Kodaira fibred surfaces weakly * rigid?

Let us explain why one can ask this question: a Kodaira fibration f:S→B yields a holomorphic map φ:B→Mg. The unramified covering B' associated to ker(ρ), ρ:π1(B)→Mapg admits then a holomorphic map φ':B'→Tg which lifts φ. If Mg were a classifying space the homomorphism ρ would determine the homotopy class of φ. A generalization of the cited theorems of Eells and Samson and of Hartmann ([147,209], see theorem 73)would then show that in each class there is a unique harmonic representative, thus proving that, fixed the complex structure on B, there is a unique holomorphic representative, if any. This might help to describe the locus thus obtained in Tb (the set of maps φ':B'→Tg that are ρ-equivariant).

But indeed, our knowledge of Kodaira fibrations is still scanty, for instance the following questions are still open.

Question 141 *(1)* Given an exact sequence $1\to\pi g\to\pi\to\pi b\to 1$ where $g\geq 3$, $b\geq 2$, does there exist a Kodaira fibred surface S with fundamental group π?

(2)(Le Brun's question, see [109]) : are there Kodaira fibred surfaces with slope $c21/c2>8/3$?

(3)Are there surfaces admitting three different Kodaira fibrations?

A brief comment on question (1) above. The given exact sequence determines, via conjugation of lifts, a homomorphism $\rho{:}\pi b\to \mathrm{Out}^{+}(\pi g)=Mapg$. If the image G of ρ is finite, then π is the fundamental group of a surface isogenous to a product, and the answer is positive, as we previously explained.

Case (II) : ρ is injective, hence (as πb has no nontrivial elements of finite order) $G{=}Im(\rho)$ acts freely on Teichmüller space Tg, and $M{:=}\mathrm{T}g/G$ is a classifying space for πb. There is a differentiable map of a curve of genus b into M, and the question is whether one can deform it to obtain a harmonic or holomorphic map.

Case (III) : ρ is not injective. This case is the most frequent one, since, given a Kodaira fibred surface $f{:}S\to B$, for each surjection $F{:}B'\to B$, the pull-back $S'{:=}S{\times}BB'$ is again a Kodaira fibration and ρ' factors as the composition of the surjection $\pi 1(B')\to\pi 1(B)$ with $\rho{:}\pi 1(B)\to Mapg$.

It is interesting to observe, via an elementary calculation, that the slope $K2S/e(S)>2$ can only decrease for a branched covering $B'\to B$, and that it tends to 2 if moreover the weight of the branch divisor $\beta{:=}\sum(1-1/mi)$tends to ∞.

Similar phenomena ($ker(\rho)\neq 0$, and the slope decreases) occur if we take general hypersurface sections of large degree $B\subset \mathrm{M}{*}g$ in the Satake compactification of the moduli space Mg which intersect neither the boundary ∂ $\mathrm{M}{*}g{=}\mathrm{M}{*}g\backslash \mathrm{M}g$ nor the locus of curves with automorphisms (both have codimension at least 2).

7.9 Can we work with locally symmetric varieties?

As we shall see in the sequel, Abelian varieties and curves are simpler objects to work with, because explicit constructions may be performed via bilinear algebra in the former case, and via ramified coverings in the latter. For locally symmetric varieties, the constructions are more difficult, hence rigidity results are not enough to decide in concrete cases whether a given variety is locally symmetric.

Recently, as a consequence of the theorem of Aubin and Yau on the existence of a Kähler–Einstein metric on projective varieties with ample canonical divisor K_X, there have been several simple explicit criteria which guarantee that a projective variety with ample canonical divisor K_X is locally symmetric (see [117, 118, 362,377]).

A difficult question which remains open is the one of studying actions of a finite group G on them, and especially of describing the G-invariant divisors.

8 INOUE TYPE VARIETIES

While a couple of hundreds examples are known today of families of minimal surfaces of general type with geometric genus $p_g(S):=\dim H^0(\mathcal{O}_S(K_S))=0$ (observe that for these surfaces $1\leq K_S^2\leq 9$), for the value $K_S^2=7$ there are only two examples known (cf. [121, 222]), and for a long time only one family of such surfaces was known, the one constructed by Masahisa Inoue (cf. [222]).

The attempt to prove that Inoue surfaces form a connected component of the moduli space of surfaces of general type proved to be successful [35], and was based on a weak rigidity result: the topological type of an Inoue surface determines an irreducible connected component of the moduli space (a phenomenon similar to the one which was observed in several papers, as [30, 31, 36, 119]).

The starting point was the calculation of the fundamental group of an Inoue surface with $p_g=0$ and $K_S^2=7$: it sits in an extension (π_g being as usual the fundamental group of a projective curve of genus g):

$$1 \to \pi_5 \times \mathbb{Z}^4 \to \pi_1(S) \to (\mathbb{Z}/2\mathbb{Z})^5 \to 1.$$

This extension is given geometrically, i.e., stems from the observation [35] that an Inoue surface S admits an unramified $(\mathbb{Z}/2\mathbb{Z})^5$—Galois covering $\hat{S}$ which is an ample divisor in $E_1\times E_2\times D$, where E_1,E_2 are elliptic curves and D is a projective curve of genus 5; while Inoue described $\hat{S}$ as a complete intersection of two non ample divisors in the product $E_1\times E_2\times E_3\times E_4$ of four elliptic curves.

It turned out that the ideas needed to treat this special family of Inoue surfaces could be put in a rather general framework, valid in all dimensions, setting then the stage for the investigation and search for a new class of varieties, which we proposed to call Inoue-type varieties.

Definition 142 ([35]) Define a complex projective manifold *X* to be an **Inoue-type manifold** if

(1)$dim(X) \geq 2$;

(2)there is a finite group *G* and an unramified *G*-covering $\hat{X} \to X$, (hence $X = \hat{X}/G$) such that

(3)$\hat{X}$ is an ample divisor inside a $K(\Gamma,1)$-projective manifold *Z*, (hence by the theorems of Lefschetz, see Theorem 3, $\pi_1(\hat{X}) \cong \pi_1(Z) \cong \Gamma$) and moreover

(4)the action of *G* on $\hat{X}$ yields a faithful action on $\pi_1(\hat{X}) \cong \Gamma$: in other words the exact sequence

$$1 \to \Gamma \cong \pi_1(\hat{X}) \to \pi_1(X) \to G \to 1$$

gives an injection $G \to \mathrm{Out}(\Gamma)$, defined by conjugation by lifts of elements of *G*;

(5)the action of *G* on $\hat{X}$ is induced by an action of *G* on *Z*.

Similarly one defines the notion of an **Inoue-type variety**, by requiring the same properties for a variety *X*with canonical singularities.

Example 143 Indeed, the examples of Inoue, which also allow him to find another description of the surfaces constructed by Burniat [65], are based on products of curves, and there the group *G* is a Z/2-vector space.

In fact things can be done more algebraically (as in [36]). If we take a $(\mathbb{Z}/2)^3$-covering of $\mathbb{P}^1$ branched on 4 points, it has equations in $\mathbb{P}^3$:

$$x_1^2 + x_2^2 + x_3^2 = 0, \quad x_0^2 - a_2 x_2^2 - a_3 x_3^2 = 0,$$

the group $G \cong (\mathbb{Z}/2)^3$ is the group of transformations $x_i \mapsto \pm x_i$, the quotient is

$$\mathbb{P}^1 = \left\{ y \in \mathbb{P}^3 | y_1 + y_2 + y_3 = 0, \ y_0 - a_2 y_2 - a_3 y_3 = 0 \right\},$$

and the branch points are the 4 points $y_i=0, i=0,1,2,3$.

The group G, if we see the elliptic curve as $\mathbb{C}/(\mathbb{Z}+\mathbb{Z}\tau)$, is the group of affine transformations

$$[z] \mapsto \pm[z] + \frac{1}{2}(a + b\tau), \quad a, b \in \{0, 1\},$$

with linear coefficient ±1 and translation vector a point of 2-torsion.

Algebra becomes easier than theta functions if one takes several square (or cubic) roots: for instance 3 square roots define a curve of genus 5 in $\mathbb{P}^4$:

$$x_1^2 + x_2^2 + x_3^2 = 0, \quad x_0^2 - a_2 x_2^2 - a_3 x_3^2 = 0, \quad x_4^2 - b_2 x_2^2 - b_3 x_3^2 = 0.$$

One can take more generally a $(\mathbb{Z}/2)^n$ covering of $\mathbb{P}^1$ branched on $n+1$ points, which is a curve of genus $g=2^{n-2}(n-3)+1$, or curves corresponding to similar Kummer coverings ($(\mathbb{Z}/m)^n$ coverings of $\mathbb{P}^1$ branched on $n+1$ points).

The above definition of Inoue type manifold, although imposing a strong restriction on X, is too general, and in order to get weak rigidity type results it is convenient to impose restrictions on the fundamental group Γ of Z, for instance the most interesting case is the one where Z is a product of Abelian varieties, curves, and other locally symmetric varieties with ample canonical bundle.

Definition 144 We shall say that an Inoue-type manifold X is

(1) a **special Inoue type manifold** if moreover

$Z=(A_1\times\cdots\times A_r)\times(C_1\times\cdots\times C_h)\times(M_1\times\cdots\times M_s)$

where each A_i is an Abelian variety, each C_j is a curve of genus $g_j\geq 2$, and M_i is a compact quotient of an irreducible bounded symmetric domain of dimension at least 2 by a torsion free subgroup;

(2) a **classical Inoue type manifold** if moreover $Z=(A_1\times\cdots\times A_r)\times(C_1\times\cdots\times C_h)$ where each A_i is an Abelian variety, each C_j is a curve of genus $g_j\geq 2$;

(3) a special Inoue type manifold is said to be **diagonal** if moreover:

the action of G on $\hat{X}$ is induced by a diagonal action on Z, i.e.,

$$G \subset \prod_{i=1}^{r} \mathrm{Aut}(A_i) \times \prod_{j=1}^{h} \mathrm{Aut}(C_j) \times \prod_{l=1}^{s} \mathrm{Aut}(M_l) \tag{145}$$

and furthermore:

(II)the faithful action on $\pi 1(\hat{}X)\cong\Gamma$, induced by conjugation by lifts of elements of *G* in the exact sequence

$$1 \to \Gamma = \Pi_{i=1}^{r}(\Lambda_i) \times \Pi_{j=1}^{h}(\pi_{g_j}) \times \Pi_{l=1}^{s}(\pi_1(M_l)) \to \pi_1(X) \to G \to 1 \tag{146}$$

(observe that each factor Λi, resp. πgj,$\pi 1(Ml)$ is a normal subgroup), satisfies the **Schur property**

$$(SP) \quad \mathrm{Hom}(V_i, V_j)^G = 0, \quad \forall i \neq j.$$

Here $Vj:=\Lambda j\otimes \mathbb{Q}$ and, in order that the Schur property holds, it suffices for instance to verify that for each Λi there is a subgroup Hi of G for which Hom $(V_i, V_j)^{H_i} = 0, \forall j \neq i.$.

The Schur property (SP) plays an important role in order to show that an Abelian variety with such a *G*-action on its fundamental group must split as a product.

Before stating the main general result of [35] we need the following definition, which was already used in 77for the characterization of complex tori among Kähler manifolds.

Definition 147 Let *Y*, *Y′* be two projective manifolds with isomorphic fundamental groups. We identify the respective fundamental groups $\pi 1(Y)=\pi 1(Y')=\Gamma$. Then we say that the condition **(SAME HOMOLOGY)** is satisfied for*Y* and *Y′* if there is an isomorphism $\Psi:H*(Y',Z)\cong H*(Y,Z)$ of homology groups which is compatible with the homomorphisms

$$u\colon H_*(Y, \mathbb{Z}) \to H_*(\Gamma, \mathbb{Z}), \quad u'\colon H_*(Y', \mathbb{Z}) \to H_*(\Gamma, \mathbb{Z}),$$

i.e., Ψ satisfies $u\circ\Psi=u'$.

We can now state the following

Theorem 148 Let *X* be a diagonal special Inoue type manifold, and let *X′* be a projective manifold with *KX′* nef and with the same fundamental group as *X*, which moreover either

(A) is homotopically equivalent to *X*;

or satisfies the following weaker property:

(B) let $\hat{X}'$ be the corresponding unramified covering of X'. Then $\hat{X}$ and $\hat{X}'$ satisfy the condition **(SAME HOMOLOGY)**.

Setting $W := \hat{X}'$, we have that

(1) $X' = W/G$ where W admits a generically finite morphism $f: W \to Z'$, and where Z' is also a $K(\Gamma,1)$projective manifold, of the form $Z' = (A'_1 \times \cdots \times A'_r) \times (C'_1 \times \cdots \times C'_h) \times (M'_1 \times \cdots \times M'_s)$.

Moreover here M'_i is either M_i or its complex conjugate, and the product decomposition corresponds to the product decomposition (146) of the fundamental group of Z.

The image cohomology class $f_*([W])$ corresponds, up to sign, to the cohomology class of $\hat{X}$.

(2) The morphism f is finite if $n = \dim X$ is odd, and it is generically injective if

(**) the cohomology class of $\hat{X}$ (in $H^*(Z,\mathbb{Z})$) is indivisible, or if every strictly submultiple cohomology class cannot be represented by an effective G-invariant divisor on any pair (Z',G) homotopically equivalent to (Z, G).

(3) f is an embedding if moreover $K_{X'}$ is ample,

(*) every such divisor W of Z' is ample, and

(***) $K^n_{X'} = K^n_{X}$.[10]

In particular, if $K_{X'}$ is ample and (*), (**) and (***) hold, also X' is a diagonal SIT (special Inoue type) manifold.

A similar conclusion holds under the alternative assumption that the homotopy equivalence sends the canonical class of W to that of $\hat{X}$: then X' is a minimal resolution of a diagonal SIT (special Inoue type) variety.

Hypothesis (A) in Theorem 148 is used to derive the conclusion that also $W := \hat{X}'$ admits a holomorphic map f' to a complex manifold Z' with the same structure as Z, while hypotheses (B) and following ensure that the morphism is birational onto its image, and the class of the image divisor $f'(\hat{X}')$ corresponds to ± that of $\hat{X}$under the identification

$$H_*(Z', \mathbb{Z}) \cong H_*(\Gamma, \mathbb{Z}) \cong H_*(Z, \mathbb{Z}).$$

Since KX' is ample, one uses (***) to conclude that f' is an isomorphism with its image.

The next question which the theorem leaves open is weak * rigidity, for which several ingredients should come into play: the Hodge type, a fine analysis of the structure of the action of G on Z, the problem of existence of hypersurfaces on which G acts freely and the study of the family of such invariant effective divisors, in particular whether the family has a connected base.

We have restricted ourselves to special Inoue type manifolds in order to be able to use the regularity results for classifying maps discussed in the previous section (the diagonality assumption is only a simplifying assumption).

Let us now sketch the proof of Theorem 148.

***Proof of Theorem** 148* **Step 1** The first step consists in showing that $W:=\hat{}X'$ admits a holomorphic mapping to a manifold Z' of the above type $Z'=(A'1\times\cdots\times A'r)\times(C'1\times\cdots\times C'h)\times(M'1\times\cdots\times M's)$, where $M'i$ is either Mi or its complex conjugate.

First of all, by the results of Siu and others ([101, 106, 344, 345], Theorem 5.14) cited in Sect. 6, W admits a holomorphic map to a product manifold of the desired type

$$Z_2' \times Z_3' = (C_1' \times \cdots \times C_h') \times (M_1' \times \cdots \times M_s').$$

Look now at the Albanese variety Alb(W) of the Kähler manifold W, whose fundamental group is the quotient of the Abelianization of $\Gamma=\pi1(Z)$ by its torsion subgroup. Write the fundamental group of Alb(W) as the first homology group of $A\times Z2\times Z3$, i.e., as

$$H_1(\mathrm{Alb}(W)) = \Lambda \oplus H_1(Z_2, \mathbb{Z}) \oplus (H_1(Z_3, \mathbb{Z})/Torsion),$$

(Alb($Z2$) is the product of Jacobians J:=(Jac($C1$)$\times\cdots\times$Jac(Ch))).

Since however, by the universal property, Alb(W) has a holomorphic map to

$$B' := \mathrm{Alb}(Z_2') \times \mathrm{Alb}(Z_3'),$$

inducing a splitting of the lattice $H1$(Alb(W),Z)=$\Lambda\oplus H1(B'$,Z), it follows that Alb(W) splits as $A'\times B'$.

Now, we want to show that the Abelian variety A' (W is assumed to be a projective manifold) splits as desired. This is in turn a consequence of assumption (3) in Definition 144. In fact, the group G acts on the Abelian variety A' as a group of biholomorphisms, hence it acts on $\Lambda \otimes \mathbb{R}$ commuting with multiplication by $\sqrt{-1}$.. Hence multiplication by $\sqrt{-1}$. is an isomorphism of G representations, and then (3) implies that $\Lambda i \otimes \mathbb{R}$ is stable by multiplication by $\sqrt{-1}$. whence $\Lambda i \otimes \mathbb{R}$ generates a subtorus $A'i$. Finally, A' splits because Λ is the direct sum of the sublattices Λi. We are through with the proof of step 1.

Step 2 Consider now the holomorphic map $f{:}W \to Z'$. We shall show that the image $W'{:=}f(W)$ is indeed a divisor in Z'. For this we use the assumption (SAME HOMOLOGY) and, in fact, the claim is an immediate consequence of the following lemma.

Lemma 149 Assume that W is a Kähler manifold, such that

(i)there is an isomorphism of fundamental groups $\pi 1(W)=\pi 1(\hat{}X)=\Gamma$, where $\hat{}X$ is a smooth ample divisor in a $K(\Gamma,1)$ complex projective manifold Z;

(ii)there exists a holomorphic map $f{:}W \to Z'$, where Z' is another $K(\Gamma,1)$ complex manifold, such that $f{*}{:}\pi 1(W) \to \pi 1(Z')=\Gamma$ is an isomorphism, and moreover

***(iii)*(SAME HOMOLOGY)** there is an isomorphism $\Psi{:}H{*}(W,\mathbb{Z}) \cong H{*}(\hat{}X,\mathbb{Z})$ of homology groups which is compatible with the homomorphisms

$$u\colon H_*(\hat{X},\mathbb{Z}) \to H_*(\Gamma,\mathbb{Z}), \quad u'\colon H_*(W,\mathbb{Z}) \to H_*(\Gamma,\mathbb{Z}),$$

i.e., we have $u \circ \Psi = u'$.

Then f is a generically finite morphism of W into Z', and the cohomology class $f{*}([W])$ in

$$H^*(Z',\mathbb{Z}) = H^*(Z,\mathbb{Z}) = H^*(\Gamma,\mathbb{Z})$$

corresponds to ±1 the one of $\hat{}X$.

Proof of the Lemma We can identify the homology groups of W and $\hat{X}$ under $\Psi:H*(W,\mathbb{Z})\cong H*(\hat{X},\mathbb{Z})$, and then the image in the homology groups of $H*(Z',\mathbb{Z})=H*(Z,\mathbb{Z})=H*(\Gamma,\mathbb{Z})$ is the same.

We apply the above consideration to the fundamental classes of the oriented manifolds W and $\hat{X}$, which are generators of the infinite cyclic top degree homology groups $H2n(W,\mathbb{Z})$, respectively $H2n(\hat{X},\mathbb{Z})$.

This implies a fortiori that $f:W\to Z'$ is generically finite: since then the homology class $f*([W])$ (which we identify to a cohomology class by virtue of Poincaré duality) equals the class of $\hat{X}$, up to sign. ▯

Step 3 We claim that $f:W\to Z'$ is generically 1-1 onto its image W'.

Let d be the degree of $f:W\to W'$.

Then $f*([W])=d[W']$, hence if the class of $\hat{X}$ is indivisible, then obviously $d=1$.

Otherwise, observe that the divisor W' is an effective G-invariant divisor and use our assumption (**).

Step 4 Here we are going to prove that f is an embedding under the additional hypotheses that $KnX=KnX'$ and that W' is ample, as well as KX'.

We established that f is birational onto its image W', hence it is a desingularization of W'.

We use now adjunction. We claim that, since (by our assumption on KX') KW is nef, there exists an effective divisor U, called the adjunction divisor, such that

$$K_W = f^*(K_{Z'} + W') - \mathfrak{U}.$$

This can be shown by taking the Stein factorization

$$p \circ h: W \to W^N \to W',$$

where WN is the normalization of W'.

Let C be the conductor ideal $\mathcal{H}om(p*\mathcal{O}WN,\mathcal{O}W')$ viewed as an ideal $\mathcal{C}\subset\mathcal{O}WN$; then the Zariski canonical divisor of WN satisfies

$$K_{W^N} = p^*(K_{W'}) - C = p^*(K_{Z'} + W') - C$$

where C is the Weil divisor associated to the conductor ideal (the equality on the Gorenstein locus of WN is shown for instance in [83], then it suffices to take the direct image from the open set to the whole of WN).

In turn, we would have in general $KW=h*(KWN)-B$, with B not necessarily effective; but, by Lemma 2.5 of [128], see also Lemma 3.39 of [257], and since −B is h-nef, we conclude that B is effective. We establish the claim by setting U:=B+$h*C$.

Observe that, under the isomorphism of homology groups, $f*(KZ'+W')$ corresponds to $(K_Z+\hat{X})|_{\hat{X}}=K_{\hat{X}}$, in particular we have

$$K^n_{\hat{X}}=|G|K^n_X=|G|K^n_{X'}=K^n_W,$$

If we assume that KW is ample, then $(KW+U)n \geq (KW)n$, equality holding if and only if U=0.

Under assumption (**), it follows that

$Kn\hat{}X=|G|KnX=|G|KnX'=KnW,$

hence U=0. Since however KW is ample, it follows that f is an embedding.

If instead we assume that KW has the same class as $f*(KZ'+W')$, we conclude first that necessarily B=0, and then we get that C=0.

Hence W' is normal and has canonical singularities.

Step 5 Finally, the group G acts on W, preserving the direct summands of its fundamental group Γ. Hence, Gacts on the curve-factors, and on the locally symmetric factors.

By assumption, moreover, it sends the summand Λi to itself, hence we get a well defined linear action on each Abelian variety $A'i$, so that we have a diagonal linear action of G on A'.

Since however the image of W generates A', we can extend the action of G on W to a compatible affine action on A'. In the case where G is abelian11, we can show that the *real affine type* of the action on A' is uniquely determined. This is taken care of by the following lemma.

Lemma 150 Given a diagonal special Inoue type manifold, if G is abelian, the real affine type of the action of G on the Abelian variety A=($A$1×⋯×Ar) is determined by the fundamental group exact sequence

$$1 \to \Gamma = \Pi_{i=1}^{r}(\Lambda_i) \times \Pi_{j=1}^{h}(\pi_{g_j}) \times \Pi_{l=1}^{s}(\pi_1(M_l)) \to \pi_1(X) \to G \to 1.$$

Proof Define as before Λ:=Πri=1Λi=π1(A); moreover, since all the summands on the left hand side are normal in π1(X), set

$$\overline{G} := \pi_1(X)/((\Pi_{j=1}^{h}\pi_{g_j}) \times (\Pi_{l=1}^{s}\pi_1(M_l))).$$

Observe now that X is the quotient of its universal covering

$$\tilde{X} = \mathbb{C}^m \times \prod_{j=1}^{h} \mathbb{H}_j \times \prod_{l=1}^{s} \mathcal{D}_l$$

by its fundamental group, acting diagonally (here Hj is a copy of Poincaré's upper half plane, while Dl is an irreducible bounded symmetric domain of dimension at least two), hence we obtain that ¯G acts on Cm as a group of affine transformations.

This action yields a homomorphism

$$\alpha : \overline{G} \twoheadrightarrow \mathrm{Im}(\alpha) =: \hat{G} \subset \mathrm{Aff}(m, \mathbb{C}) :$$

let K the kernel of α, and let

$$\overline{G_1} := \ker(\alpha_L : \overline{G} \to \mathrm{GL}(m, \mathbb{C})).$$

$\overline{G_1}$ is obviously Abelian, and contains Λ, and maps onto a lattice Λ′⊂ˆG.

Since Λ injects into Λ′, Λ∩K=0, whence K injects into G, therefore K is a torsion subgroup; since Λ′ is free, we obtain

$$\overline{G_1} = \Lambda' \oplus \mathcal{K},$$

and we finally get

$$\mathcal{K} = \mathrm{Tors}(\overline{G_1}), \quad \hat{G} = \overline{G}/\,\mathrm{Tors}(\overline{G_1}).$$

Since our action is diagonal, we can write $\Lambda' = \oplus_{i=1}^{r} \Lambda'_i$, and the linear action of the group $G2{:=}G/K$preserves the summands.

Since $\hat{G} \subset \mathrm{Aff}(\Lambda' \otimes \mathbb{R})$, we can now apply Proposition 21 to it.

This shows that the affine group $\hat{G}$ is uniquely determined.

Finally, using the image groups $G2,i$ of $G2$ inside GL($\Lambda'i$), we can define uniquely groups of affine transformations of Ai which fully determine the diagonal action of G on A (up to real affine automorphisms of each Ai).

The proof of Theorem 148 is now completed. ▢

In order to obtain weak rigidity results, one has to use, as an invariant for group actions on tori and Abelian varieties, the Hodge type, introduced in Definition 19 (see also Remark 20).

Remark 151 In the previous theorem special assumptions are needed in order to guarantee that for each manifod X'homotopy equivalent to X the classifying holomorphic map $f : \hat{X}' \to Z'$ be birational onto its image, and indeed an embedding.

However, there is the possibility that an Inoue type variety deforms to one for which f is a covering of finite degree. This situation should be analyzed and the singularities of the image of f described in detail, so as to lead to a generalization of the theory of Inoue type varieties, including 'multiple' Inoue type varieties (those for which f has degree at least two).

The study of moduli spaces of Inoue type varieties, and their connected and irreducible components, relies very much on the study of moduli spaces of varieties X endowed with the action of a finite group G: and it is for us a strong motivation to pursue this line of research.

This topic will occupy a central role in the following sections, first in general, and then in the special case of algebraic curves.

9 MODULI SPACES OF SURFACES AND HIGHER DIMENSIONAL VARIETIES

Teichmüller theory works out quite fine in the case of projective curves, as well as other approaches, like Geometric Invariant Theory (see [178, 179, 302, 304]), which provides a quasi-projective moduli space Mgendowed with a natural compactification $\overline{\mathfrak{M}_g}$ (this is a coarse moduli space for the so-called moduli stable projective curves: these are the reduced curves with at most nodes as singularities, such that their group of automorphisms is finite).

In higher dimensions one has a fully satisfactory theory of 'local moduli' for compact complex manifolds or spaces, but there are difficulties with the global theory.

So, let us start from the local theory, developed by Kodaira and Spencer, and culminating in the results of Kuranishi and Grauert.

9.1 Kodaira–Spencer–Kuranishi theory

While describing complex structures as integrable almost complex structures (see Theorem 32), it is convenient to view an almost complex structure as a differentiable (0, 1)-form with values in the dual of the cotangent bundle $(TY1,0)\vee$.

This representation leads to the Kodaira–Spencer–Kuranishi theory of local deformations, addressing precisely the study of the **small deformations** of a complex manifold $Y=(M,J0)$.

In this theory, complex structures correspond to closed such forms which, by Dolbeault's theorem, determine a cohomology class in $H1(\Theta Y)$, where ΘY is the sheaf of holomorphic sections of the holomorphic tangent bundle $TY1,0$.

We shall use here unambiguously the double notation $TM0,1=TY0,1$, $TM1,0=TY1,0$ to refer to the splitting of the complexified tangent bundle determined by the complex structure $J0$.

$J0$ is a point in $C(M)$, and a neighbourhood in the space of almost complex structures corresponds to a distribution of subspaces which are globally defined as graphs of an endomorphism

$$\phi : TM^{0,1} \to TM^{1,0},$$

called a **small variation of almost complex structure**, since one then defines

$$TM_\phi^{0,1} := \{(u, \phi(u)) | \ u \in TM^{0,1}\} \subset TM^{0,1} \oplus TM^{1,0}.$$

In terms of the old $\bar{\partial}$ operator, the new one is simply obtained by considering

$$\bar{\partial}_\phi := \bar{\partial} + \phi,$$

and the integrability condition is given by the Maurer–Cartan equation

$$(MC) \ \bar{\partial}(\phi) + \frac{1}{2}[\phi, \phi] = 0.$$

Observe that, since our original complex structure $J0$ corresponds to ϕ=0, the derivative of the above equation at ϕ=0 is simply

$$\bar{\partial}(\phi) = 0,$$

hence the tangent space to the space of complex structures consists of the space of $\bar{\partial}$-closed forms of type (0, 1) and with values in the bundle $TM^{1,0}$.

One can restrict oneself (see e.g. [114]) to consider only the class of such forms ϕ in the Dolbeault cohomology group

$$H^1(\Theta_Y) := ker(\bar{\partial})/Im(\bar{\partial}),$$

by looking at the action of the group of diffeomorphisms which are exponentials of global vector fields on M.

Representing these cohomology classes by harmonic forms, the integrability equation becomes easier to solve via the following Kuranishi equation.

Let $\eta 1,\dots,\eta m \in H1(\Theta Y)$ be a basis for the space of harmonic (0,1)-forms with values in $TM1,0$, and set $t:=(t1,\dots,tm)\in Cm$, so that $t \mapsto \sum i t i \eta i$ establishes an isomorphism $Cm \cong H1(\Theta Y)$.

Then the *Kuranishi slice* is obtained by associating to *t* the unique power series solution of the following equation:

$$\phi(t) = \sum_i t_i \eta_i + \frac{1}{2}\bar{\partial}^* G[\phi(t), \phi(t)],$$

satisfying moreover $\phi(t)=\sum_i t_i\eta_i+$ higher order terms (*G* denotes here the Green operator).

The upshot is that for these forms $\phi(t)$ the integrability equation simplifies drastically; the result is summarized in the following definition.

Definition 152 The Kuranishi space B(*Y*) is defined as the germ of complex subspace of $H^1(\Theta_Y)$ defined by $\{t\in\mathbb{C}^m|\ H[\phi(t),\phi(t)]=0\}$, where *H* is the harmonic projector onto the space $H^2(\Theta_Y)$ of harmonic forms of type (0, 2) and with values in $TM^{1,0}$.

Kuranishi space B(*Y*) parametrizes the set of small variations of almost complex structure $\phi(t)$ which are integrable. Hence over B(*Y*) we have a family of complex structures which deform the complex structure of *Y*.

It follows then that the Kuranishi space B(*Y*) surjects onto the germ of the Teichmüller space at the point corresponding to the given complex structure $Y=(M,J_0)$.

It fails badly in general to be a homeomorphism (see [86, 114]): for instance Teichmüller space, in the case of the Hirzebruch Segre surface F_{2n} (blow up of the cone onto the rational normal curve of degree 2*n*), consists of *n*+1 points $p_0,p_2,\ldots,p_{2n}$, such that the open sets are exactly the sets $\{p_{2i}|i\le k\}$.

But a first consequence is that Teichmüller space is locally connected by holomorphic arcs, hence the determination of the connected components of C(*M*), respectively of T(*M*), can be done using the original definition of deformation equivalence, given by Kodaira and Spencer [248] (see Definition 109).

One can define deformations not only for complex manifolds, but also for complex spaces.

Definition 153 (1) A **deformation** of a compact complex space *X* is a pair consisting of

(1.1) a flat proper morphism $\pi:\mathcal{X}\to T$ between connected complex spaces (i.e., $\pi^*:\mathcal{O}_{T,t}\to\mathcal{O}_{\mathcal{X},x}$ is a flat ring extension for each *x* with $\pi(x)=t$)

(1.2) an isomorphism $\psi: X \cong \pi^{-1}(t_0) := X_0$ of X with a fibre X_0 of π.

(2.1) A **small deformation** is the germ $\pi: (\mathcal{X}, X_0) \to (T, t_0)$ of a deformation.

(2.2) Given a deformation $\pi: \mathcal{X} \to T$ and a morphism $f: T' \to T$ with $f(t'_0) = t_0$, the **pull-back** $f^*(\mathcal{X})$ is the fibre product $\mathcal{X}' := \mathcal{X} \times_T T'$ endowed with the projection onto the second factor T' (then $X \cong X'_0$).

(3.1) A small deformation $\pi: \mathcal{X} \to T$ is said to be **versal** or **complete** if every small deformation $\pi: \mathcal{X}' \to T'$ is obtained from it via pull back; it is said to be **semi-universal** if the differential of $f: T' \to T$ at t'_0 is uniquely determined, and **universal** if the morphism f is uniquely determined.

(4) Two compact complex manifolds X, Y are said to be **directly deformation equivalent** if there are

(4i) a deformation $\pi: \mathcal{X} \to T$ of X with T irreducible and where all the fibres are smooth, and

(4ii) an isomorphism $\psi': Y \cong \pi^{-1}(t_1) =: X_1$ of Y with a fibre X_1 of π.

Remark 154 The technical assumption of flatness replaces, for families of spaces, the condition on π to be a submersion, necessary in order that the fibres be smooth manifolds.

Let's however come back to the case of complex manifolds, observing that in a small deformation of a compact complex manifold one can shrink the base T and assume that all the fibres are smooth.

We can now state the results of Kuranishi and Wavrik [259, 260, 371] in the language of deformation theory.

Theorem 155 (Kuranishi) Let Y be a compact complex manifold: then

(I)the Kuranishi family $\pi: (\mathcal{Y}, Y_0) \to (\mathfrak{B}(Y), 0)$ of Y is semiuniversal.

(II)$(\mathfrak{B}(Y), 0)$ is unique up to (non canonical) isomorphism, and is a germ of analytic subspace of the vector space $H^1(Y, \Theta_Y)$, inverse image of the origin under a local holomorphic map (called Kuranishi map)

$$k : H^1(Y, \Theta_Y) \to H^2(Y, \Theta_Y)$$

whose differential vanishes at the origin. Moreover the quadratic term in the Taylor development of k is given by the bilinear map

$H1(Y,\Theta Y) \times H1(Y,\Theta Y) \to H2(Y,\Theta Y)$, called Schouten bracket, which is the composition of cup product followed by Lie bracket of vector fields.

(III)The Kuranishi family is a versal deformation of Yt for $t \in B(Y)$.

(IV)The Kuranishi family is universal if $H0(Y,\Theta Y)=0$.

***(V)*(Wavrik)** The Kuranishi family is universal if $B(Y)$ is reduced and $h0(Yt,\Theta Yt)$:=dim $H0(Yt,\Theta Yt)$ is constant for $t \in B(Y)$ in a suitable neighbourhood of 0.

The Kodaira Spencer map, defined soon below, is to be thought as the derivative of a family of complex structures.

Definition 156 The Kodaira Spencer map of a family π:(Y,$Y0$)→(T,$t0$) of complex manifolds having a smooth base T is defined as follows: consider the cotangent bundle sequence of the fibration

$$0 \to \pi^*(\Omega^1_T) \to \Omega^1_{\mathcal{Y}} \to \Omega^1_{\mathcal{Y}|T} \to 0,$$

and the direct image sequence of the dual sequence of bundles,

$$0 \to \pi_*(\Theta_{\mathcal{Y}|T}) \to \pi_*(\Theta_{\mathcal{Y}}) \to \Theta_T \to \mathcal{R}^1\pi_*(\Theta_{\mathcal{Y}|T}).$$

Evaluation at the point $t0$ yields a map ρ of the tangent space to T at $t0$ into $H1(Y0,\Theta Y0)$, which is the derivative of the variation of complex structure.

The Kodaira Spencer map and the implicit functions theorem allow to determine the Kuranishi space and the Kuranishi family in many cases.

Corollary 157 Let Y be a compact complex manifold and assume that we have a family π:(Y,$Y0$)→(T,$t0$) with smooth base$T$, such that $Y \cong Y0$, and such that the Kodaira Spencer map $\rho t0$ surjects onto $H1(Y,\Theta Y)$.

Then the Kuranishi space $B(Y)$ is smooth and there is a submanifold $T' \subset T$ which maps isomorphically to $B(Y)$; hence the Kuranishi family is the restriction of π to T'.

The key point is that, by versality of the Kuranishi family, there is a morphism f:$T \to \mathfrak{B}(Y)$ inducing π as a pull back, and ρ is the derivative of f: then one uses the implicit functions theorem.

This approach clearly works only if Y is **unobstructed**, which simply means that $\mathfrak{B}(Y)$ is smooth. In general it is difficult to describe the obstruction map, and even calculating the quadratic term is nontrivial (see [216] for an interesting example).

Even if it is difficult to calculate the obstruction map, Kuranishi theory gives a lower bound for the 'number of moduli' of Y, since it shows that $\mathfrak{B}(Y)$ has dimension $\geq h^1(Y,\Theta_Y)-h^2(Y,\Theta_Y)$. In the case of curves $H^2(Y,\Theta_Y)=0$, hence curves are unobstructed; in the case of a surface S

$$\dim \mathfrak{B}(S) \geq h^1(\Theta_S) - h^2(\Theta_S) = -\chi(\Theta_S) + h^0(\Theta_S) = 10\chi(\mathcal{O}_S) - 2K_S^2 + h^0(\Theta_S).$$

The above is the Enriques inequality ([157], see also [155], and [126] for an improvement: observe that Max Noether postulated equality), proved by Kuranishi in all cases and also for non algebraic surfaces.

9.2 Kuranishi and Teichmüller

Ideally, we would like to have that Teichmüller space, up to now only defined as a topological space, is indeed a complex space, locally isomorphic to Kuranishi space.

In fact, we already remarked that there is a locally surjective continuous map of $\mathcal{B}(Y)$ to the germ $\mathcal{T}(M)_Y$ of $\mathcal{T}(M)$ at the point corresponding to the complex structure yielding Y. For curves this map is a local homeomorphism, and this fact provides a complex structure on Teichmüller space.

Whether this holds in general is related to the following definition.

Definition 158 A compact complex manifold Y is said to be **rigidified** if $Aut(Y)\cap \mathrm{Diff}^0(Y)=\{Id_Y\}$.

A compact complex manifold Y is said to be cohomologically rigidified if $Aut(Y)\to Aut(H^*(Y,\mathbb{Z}))$ is injective, and rationally cohomologically rigidified if $Aut(Y)\to Aut(H^*(Y,\mathbb{Q}))$ is injective.

Remark 159 We refer to [68] for the proof that surfaces of general type with $q\geq 3$ are cohomologically rigidified, and for examples showing that there are rigidified surfaces of general type which are not cohomologically rigidified.

In fact, it is clear that there is a universal tautological family of complex structures parametrized by $\mathcal{C}(M)$ (the closed subspace $\mathcal{C}(M)$ of $\mathcal{AC}(M)$ consisting of the set of complex structures on M), and with total space

$$\mathfrak{U}_{\mathcal{C}(M)} := M \times \mathcal{C}(M),$$

on which the group $Diff^{+}(M)$ naturally acts, in particular $Diff^{0}(M)$.

The main observation is that $Diff^{0}(M)$ acts freely on C(M) if and only if for each complex structure Y on M the group of biholomorphisms $Aut(Y)$ contains no nontrivial automorphism which is differentiably isotopic to the identity. Thus, the condition of being rigidified implies that the tautological family of complex structures descends to a universal family of complex structures on Teichmüller space:

$$\mathfrak{U}_{\mathcal{T}(M)} := (M \times \mathcal{C}(M))/\mathcal{D}iff^{0}(M) \to \mathcal{C}(M))/\mathcal{D}iff^{0}(M) = \mathcal{T}(M),$$

on which the mapping class group acts.

Fix now a complex structure yielding a compact complex manifold Y, and compare with the Kuranishi family

$$\mathcal{Y} \to \mathfrak{B}(Y).$$

Now, we already remarked that there is a locally surjective continuous map of B(Y) to the germ T(M)Y of T(M)at the point corresponding to the complex structure yielding Y.

The following was observed in [114].

Remark 160 If

(1)the Kuranishi family is universal at any point

(2) $\mathfrak{B}(Y) \to \mathcal{T}(M)_Y$is injective (it is then a local homeomorphism at every point)

then Teichmüller space has a natural structure of complex space.

In many cases (for instance, complex tori) Kuranishi and Teichmüller space coincide, in spite of the fact that the manifolds are not rigidified. For instance we showed in [114]:

Proposition 161 *(1)* The continuous map π:B(Y)→T(M)Y is a local homeomorphism between Kuranishi space and Teichmüller space if there is an injective continuous map f:B(Y)→Z, where Z is Hausdorff, which factors through π.

(2)Assume that Y is a compact Kähler manifold and that the local period map f is injective: then π:B(Y)→T(M)Y is a local homeomorphism.

*(3)*In particular, this holds if Y is Kähler with trivial or torsion canonical divisor.

Remark 162(1)The condition of being rigidified implies the condition $H0(\Theta Y)=0$ (else there is a positive dimensional Lie group of biholomorphic self maps), and is obviously implied by the condition of being cohomologically rigidified.

*(2)*By the cited Lefschetz' lemma compact curves of genus $g\geq 2$ are rationally cohomologically rigidified, and it is an interesting question whether compact complex manifolds of general type are rigidified (see [67] and [68] for recent progresses).

*(3)*One can dispense with many assumptions, at the cost of having a more complicated result. For instance, Meersseman shows in [284], explaining the situation via classical examples of [81], that one can take a quotient of the several Kuranishi families (using their semi-universality), obtaining, as analogues of Teichmüller spaces, respectively moduli spaces, an analytic Teichmüller groupoid, respectively a Riemann groupoid, both independent up to analytic Morita equivalence of the chosen countable disjoint union of Kuranishi spaces. These lead to some stacks called by the author 'analytic Artin stack' (see [11] and [164] as a reference on Artin stacks).

9.3 Varieties with singularities

For higher dimensional varieties moduli theory works better if one considers varieties with moderate singularities, rather than smooth ones (see [256]), as we shall illustrate in the next sections for the case of algebraic surfaces.

The Kuranishi theory does indeed extend perfectly to all compact complex spaces, and Kuranishi's theorem was generalized by Grauert (see [186, 187], see also [334] for the algebraic analogue).

Theorem 163 Grauert's Kuranishi type theorem for complex spaces. Let X be a compact complex space: then

*(I)*there is a semiuniversal deformation π:(X,$X0$)→(T,$t0$) of X, i.e., a deformation such that every small deformation π':(X′,$X'0$)→(T',$t'0$) is the pull-back of π for an appropriate morphism f:(T',$t'0$)→(T,$t0$)whose differential at $t'0$ is uniquely determined.

(II)$(T,t0)$ is unique up to isomorphism, and is a germ of analytic subspace of the vector space T1 of first order deformations.

$(T,t0)$ is the inverse image of the origin under a local holomorphic map (called Kuranishi map)

k:$T^1 \to T2$

to the finite dimensional vector space T2 (called **obstruction space**), and whose differential vanishes at the origin (the point corresponding to the point $t0$).

If X is reduced, or if the singularities of X are local complete intersection singularities, then $\mathbb{T}^1 = \mathrm{Ext}^1(\Omega^1_X, \mathcal{O}_X)$.

If the singularities of X are local complete intersection singularities, then T2=$\mathrm{Ext}^2(\Omega^1_X, \mathcal{O}_X)$.

Indeed, the singularities which occur for the canonical models of varieties of general type are called canonical singularities and are somehow tractable (see [326]); nevertheless the study of their deformations may present highly nontrivial problems.

For this reason, we restrict now ourselves to the case of complex dimension two, where these singularities are easier to describe.

10 MODULI SPACES OF SURFACES OF GENERAL TYPE

10.1 Canonical models of surfaces of general type.

The classification theory of algebraic varieties proposes to classify the birational equivalence classes of projective varieties.

Now, in the birational class of a non ruled projective surface there is, by the theorem of Castelnuovo (see e.g. [55]), a unique (up to isomorphism) minimal model S (concretely, minimal means that S contains no (−1)-curves, i.e., curves E such that $E \cong$P1, and with $E^2=-1$).

We shall assume from now on that S is a smooth minimal (projective) surface of general type: this is equivalent (see [55]) to the two conditions:

(*) $K_S^2 > 0$ and KS is nef, where as well known, a divisor D is said to be nef if, for each irreducible curve C, we have $D \cdot C \geq 0$.

It is very important that, as shown by Kodaira [250], the class of non minimal surfaces is stable by small deformation; on the other hand, a small deformation of a minimal algebraic surface of general type is again minimal (see prop. 5.5 of [21]). Therefore, the class of minimal algebraic surfaces of general type is stable by deformation in the large.

Even if the canonical divisor KS is nef, it does not however need to be an ample divisor, indeed

The canonical divisor KS of a minimal surface of general type S is ample iff there does not exist an irreducible curve C on S with $K \cdot C = 0 \Leftrightarrow$ *there is no* (−2)-*curve C on S, i.e., a curve such that* $C \cong$ P1, *and* $C^2 = -2$.

The number of (−2)-curves is bounded by the rank of the Neron Severi lattice $NS(S) \subset H2(S,Z)/Torsion$ of $S(NS(S)$ is the image of $Pic(S) = H^1(O*S)$ inside $H^2(S,Z)/Torsion$, and, by Lefschetz' (1,1) theorem, it is the intersection with the Hodge summand $H1,1)$; the (−2)-curves can be contracted by a contraction $\pi: S \to X$, where X is a normal surface which is called the **canonical model** of S.

The singularities of X are called Rational Double Points (also called Du Val or Kleinian singularities, see [142]), and X is a Gorenstein variety , i.e. (see [211]) the dualizing sheaf ωX is invertible, and the associated Cartier divisor KX, called again canonical divisor, is ample and such that $\pi*(KX) = KS$.

X is also the projective spectrum (set of homogeneous prime ideals) of the canonical ring

$$\mathcal{R}(S) := \mathcal{R}(S, K_S) := \bigoplus_{m \geq 0} H^0(\mathcal{O}_S(mK_S)).$$

This definition generalizes to any dimension, since for a variety of general type (one for which there is a pluricanonical map which is birational onto its image) the canonical ring is a finitely generated graded ring, as was proven by Birkar, Cascini, Hacon, and McKernan [43]. And one defines a canonical model as a variety with canonical singularities and with ample canonical divisor KX.

More concretely, the canonical model of a surface of general type is directly obtained as the image, for $m \geq 5$, of the m-th pluricanonical map of S (associated to the sections in $H0(OS(mKS))$) as shown by Bombieri [54] (extended to any characteristic by Ekedahl [151], and in [93, 94]).

Theorem 164 (Bombieri) Let S be a minimal surface of general type, and consider the m-th pluricanonical map φm of S(associated to the linear system $|mKS|$) for $m \geq 5$, or for $m = 4$ when $K_S^2 \geq 2$.

Then φm is a birational morphism whose image is isomorphic to the canonical model X, embedded by its m-th pluricanonical map.

Corollary 165 Bombieri [55] Minimal surfaces S of general type with given topological invariants $e(S)$,$b+(S)$ (here, $b+(S)$ is the positivity of the intersetion form on $H2(S,Z)$, and the pair of topological invariants is equivalent to the pair of holomorphic invariants $K2S$,$\chi(S):=\chi(OS)$) are 'bounded', i.e., they belong to a finite number of families having a connected base.

In particular, for fixed Euler number and positivity $e(S)$,$b+(S)$ we have a finite number of differentiable and topological types.

In fact, one can deduce from here effective upper bounds for the number of these families, hence for the types (see [88, 92]); see [273] for recent work on lower bounds and references to other recent and less recent results.

Results in the style of Bombieri have been recently obtained by Hacon and Mckernan [205], and from these one obtains (non explicit) boundedness results for the varieties of general type with fixed invariants $K_X^n, \chi(X)$.

10.2 The Gieseker moduli space

When one deals with projective varieties or projective subschemes the most natural parametrization, from the point of view of deformation theory, is given by the Hilbert scheme, introduced by Grothendieck [200].

For instance, in the case of surfaces of general type with fixed invariants $\chi(S)=a$ and $K2S=b$, their 5-canonical models $X5$ are surfaces with Rational Double Points as singularities and of degree 25 b in a fixed projective space PN, where N+1=$P5$:= $h^0(5K_S) = \chi(S) + 10K_S^2 = a + 10b$.

The Hilbert polynomial of $X5$ is the polynomial

$$P(m) := h^0(5mK_S) = a + \frac{1}{2}(5m-1)5mb.$$

Grothendieck [200] showed that, given a Hilbert polynomial (see [211]), there is

*(i)*an integer *d* and

*(ii)*a subscheme H=H*P* of the Grassmannian of codimension *P*(*d*)-subspaces of *H*0(P*N*,OP*N*(*d*)), called Hilbert scheme, such that

*(iii)*H parametrizes the degree *d* graded pieces $H^0(\mathbb{P}^N, \mathcal{O}_{\mathbb{P}^N}(d))$. of the homogeneous ideals of all the subschemes Σ⊂P*N* having the given Hilbert polynomial *P*.

The Hilbert point of Σ is the Plücker point

$$\Lambda^{P(d)}(r_\Sigma^\vee) \in \mathbb{P}(H^0(\mathbb{P}^N, \mathcal{O}_{\mathbb{P}^N}(d))^\vee)$$

where *r*Σ is the restriction homomorphism (surjective for *d* large)

$$r_\Sigma : H^0(\mathbb{P}^N, \mathcal{O}_{\mathbb{P}^N}(d)) \to H^0(\Sigma, \mathcal{O}_\Sigma(d)).$$

Inside H one has the open sets

$$\mathcal{H}^* = \{\Sigma | \Sigma \text{ is smooth}\} \subset \mathcal{H}^0,$$

where

H0:={Σ|Σ is reduced with only rational Gorenstein singularities}.

Gieseker showed in [177], replacing the 5-canonical embeddding by an *m*-canonical embedding with much higher *m*, the following

Theorem 166 (Gieseker) The moduli space of canonical models of surfaces of general type with invariants *χ*,*K*2 exists as a quasi-projective scheme

$$\mathfrak{M}^{can}_{\chi,K^2}$$

which is called the **Gieseker moduli space**.

Similar theorems hold also for higher dimensional varieties, see [256, 361], for the most recent developments, but we do not state here the results, which are more complicated and technical.

10.3 Components of moduli spaces and deformation equivalence

We mentioned previously that the relation of deformation equivalence is a good general substitute for the condition that two manifolds belong to the same connected (resp.: irreducible) component of the moduli space.

Things get rather complicated and sometimes pathological in higher dimension, especially since, even for varieties of general type, there can be three models, smooth, terminal, and canonical model, and the latter are singular.

In the case of surfaces of general type things still work out fine (up to a certain extent), since the main issue is to compare the deformations of minimal models versus the ones of canonical models.

We have then the following theorem.

Theorem 167 Given two minimal surfaces of general type S,S' and their respective canonical models X,X', then

S and S' are deformation equivalent $\Leftrightarrow$ X and X' are deformation equivalent $\Leftrightarrow$ X and X' yield two points in the same connected component of the Gieseker moduli space.

One idea behind the proof is the observation that, in order to analyse deformation equivalence, one may restrict oneself to the case of families parametrized by a base T with $dim(T)$=1: since two points in a complex space $T\subset Cn$ (or in an algebraic variety) belong to the same irreducible component of T if and only if they belong to an irreducible curve $T'\subset T$. And one may further reduce to the case where T is smooth simply by taking the normalization $T0\rightarrow Tred\rightarrow T$ of the reduction $Tred$ of T, and taking the pull-back of the family to $T0$.

A less trivial result which is used is the so-called simultaneous resolution of singularities (cf. [60–62, 353])

Theorem 168 (Simultaneous resolution according to Brieskorn and Tjurina) Let T:=Cτ be the basis of the semiuniversal deformation of a Rational Double Point (X, 0). Then there exists a ramified Galois cover $T'\to T$, with T'smooth $T'\cong$Cτ such that the pull-back X′:=X×TT' admits a simultaneous resolution of singularities p:S′→X′ (i.e., p is bimeromorphic, all the fibres of the composition S′→X′→T' are smooth, and the fibre over t'=0 is equal to the minimal resolution of singularities of (X, 0)).

Another important observation is that the local analytic structure of the Gieseker moduli space is determined by the action of the group of automorphisms of X on the Kuranishi space of X.

Remark 169 Let X be the canonical model of a minimal surface of general type S with invariants χ,$K$2. The isomorphism class of X defines a point $[X]\in\mathfrak{M}^{can}_{\chi,K^2}$.

Then the germ of complex space $(\mathfrak{M}^{can}_{\chi,K^2},[X])$ is analytically isomorphic to the quotient B(X)/Aut(X) of the Kuranishi space of X by the finite group Aut(X)=Aut(S).

Let S be a minimal surface of general type and let X be its canonical model. To avoid confusion between the corresponding Kuranishi spaces, denote by Def(S) the Kuranishi space for S, respectively Def(X) the Kuranishi space of X.

Burns and Wahl [66], inspired by [13] explained the relation holding between Def(S) and Def(X).

Theorem 170 (Burns–Wahl) Assume that KS is not ample and let π:$S\to X$ be the canonical morphism.

Denote by LX the space of local deformations of the singularities of X (Cartesian product of the corresponding Kuranishi spaces) and by LS the space of deformations of a neighbourhood of the exceptional locus $Exc(\pi)$ of π. Then Def(S) is realized as the fibre product associated to the Cartesian diagram

$$\begin{array}{ccc} Def(S) & \longrightarrow & Def(S_{Exc(\pi)}) =: \mathcal{L}_S \cong \mathbb{C}^{\nu}, \\ \Big\downarrow & & \Big\downarrow \lambda \\ Def(X) & \longrightarrow & Def(X_{Sing X}) =: \mathcal{L}_X \cong \mathbb{C}^{\nu}, \end{array}$$

where ν is the number of rational (−2)-curves in S, and λ is a Galois covering with Galois group $W:=\oplus_{i=1}^{r}W_i$, the direct sum of the Weyl groups *Wi* of the singular points of *X* (these are generated by reflections, hence yield a smooth quotient, see [124]).

An immediate consequence is the following

Corollary 171

(Burns–Wahl)

(1) $\psi : Def(S) \to Def(X)$ is a finite morphism, in particular, ψ is surjective.

(2)If the derivative of Def(*X*)→L*X* is not surjective (i.e., the singularities of *X* cannot be independently smoothened by the first order infinitesimal deformations of *X*), then Def(*S*) is singular.

In the next section we shall see the role played by automorphisms.

10.4 Automorphisms and canonical models

Assume that *G* is a group with a faithful action on a complex manifold *Y*: then *G* acts naturally on the sheaves associated to Ω1, hence we have a linear representation of *G* on the vector spaces that are the cohomology groups of such sheaves: for instance *G* acts linearly on *Hq*(Ω*pY*), and also on the vector spaces $H^0((\Omega^n_Y)^{\otimes m}) = H^0(\mathcal{O}_Y(mK_Y))$, hence on the canonical ring

$$\mathcal{R}(Y) := \mathcal{R}(Y, K_Y) := \bigoplus_{m \geq 0} H^0(\mathcal{O}_Y(mK_Y)).$$

If *Y* is a variety of general type, then *G* acts linearly on the vector space *H*0(O*Y*(*mKY*)), hence linearly on the m-th pluricanonical image *Ym*, which is an algebraic variety bimeromorphic to *Y*. Hence *G* is isomorphic to a subgroup of the algebraic group *Aut*(*Ym*). Matsumura [283] used the structure theorem for linear algebraic groups (see [218]) to show that, if *G* were infinite, then *Aut*(*Ym*) would contain a non trivial Cartan subgroup (Cor C∗), hence *Y* would be uniruled: a contradiction. Hence we have the (already mentioned) theorem:

Theorem 172 (Matsumura) The automorphism group of a variety *Y* of general type is finite.

The above considerations apply now to the *m*-th pluricanonical image *X* of a variety *Y* of general type.

I.e., we have an embedded variety $X\subset \mathbb{P}(V)$ and a linear representation of a finite group G on the vector space V, such that X is G-invariant (for Y of general type one has $V := V_m := H^0(\mathcal{O}_Y(mK_Y)))$.

Now, since we work over $\mathbb{C}$, the vector space V splits uniquely, up to permutation of the summands, as a direct sum of irreducible representations

$$(**)\ V_m = \bigoplus_{\rho\in Irr(G)} W_\rho^{n(\rho)}.$$

We come now to the basic notion of a family of G-automorphisms (this notion shall be further explained in the next section).

Definition 173 A family of G-automorphisms is a triple

$$((p\colon \mathcal{X} \to T), G, \alpha)$$

where:

(1)$(p{:}\mathrm{X}\to T)$ is a flat family in a given category (a smooth family for the case of minimal models of surfaces of general type)

(2)G is a (finite) group

(3)$\alpha{:}G\times\mathrm{X}\to\mathrm{X}$ yields a biregular action $G\to Aut(\mathrm{X})$, which is compatible with the projection p and with the trivial action of G on the base T (i.e., $p(\alpha(g,x))=p(x), \forall g\in G, x\in \mathrm{X}$).

As a shorthand notation, one may also write $g(x)$ instead of $\alpha(g,x)$, and by abuse of notation say that the family of automorphisms is a deformation of the pair (Xt,G) instead of the triple $(Xt,G,\alpha t)$.

The following then holds.

Proposition 174 *(1)* A family of automorphisms of manifolds of general type (not necessarily minimal models) induces a family of automorphisms of canonical models.

*(2)*A family of automorphisms of canonical models induces, if the basis T is connected, a constant decomposition type $(**)$ for $Vm(t)$.

*(3)*A family of automorphisms of smooth complex manifolds admits a differentiable trivialization, i.e., in a neighbourhood of $t0\in T$, a

diffeomorphism as a family with $(S0\times T, pT, \alpha 0\times IdT)$; in other words, with the trivial family for which $g(y,t)=(g(y),t)$.

We refer to [114] for a sketchy proof, let us just observe that if we have a continuous family of finite dimensional representations, the multiplicity of an irreducible summand is a locally constant function on the parameter space T (being given by the scalar product of the respective trace functions, it is an integer valued continuous function, hence locally constant).

Let us then consider the case of a family of canonical models of varieties of general type: by 2) above, and shrinking the base in order to make the addendum R(p)m=p*(OS(mK)) free, we get an embedding of the family

$$(\mathcal{X}, G) \hookrightarrow T \times \left(\mathbb{P}\left(V_m = \bigoplus_{\rho \in Irr(G)} W_\rho^{n(\rho)} \right), G \right).$$

In other words, all the canonical models Xt are contained in a fixed projective space, where also the action of$G$ is fixed.

Now, the canonical model Xt is left invariant by the action of G if and only if its Hilbert point is fixed by G. Hence, we get a closed subset $\mathcal{H}_0(\chi, K^2)^G$ of the Hilbert scheme $\mathcal{H}^0(\chi, K^2)$

$$\mathcal{H}_0(\chi, K^2)^G := \{X | \omega_X \cong \mathcal{O}_X(1), \quad X \quad is\, normal, \quad \gamma(X) = X \quad \forall \gamma \in G\}.$$

For instance in the case of surfaces one has the following theorem (see [256, 360, 361] and references therein for results on moduli spaces of canonically polarized varieties in higher dimension).

Theorem 175 The surfaces of general type which admit an action of a given pluricanonical type (**) i.e., with a fixed irreducible G-decomposition of their canonical ring, form a closed subvariety (M$can\chi$,$K$2)G,(**) of the moduli space M$can\chi$,$K$2.

Remark 176 The situation for the minimal models of surfaces of general type is different, because then the subset of the moduli space where one has a fixed differentiable type is not closed, as showed in [33].

The puzzling phenomenon which we discovered in joint work with Ingrid Bauer [32, 33], on the moduli spaces of Burniat surfaces) is that deformations of automorphisms differ for canonical and for minimal models.

More precisely, let S be a minimal surface of general type and let X be its canonical model. Denote by Def(S), resp. Def(X), the base of the Kuranishi family of S, resp. of X.

Assume now that we have $1 \neq G \leq \mathrm{Aut}(S) = \mathrm{Aut}(X)$.

Then we can consider the Kuranishi space of G-invariant deformations of S, denoted by Def(S,G), and respectively consider Def(X,G); we have a natural map Def(S,G)→Def(X,G).

We indeed show in [33] that this map needs not be surjective, even if surjectivity would seem plausible; we have the following result:

Theorem 177 The deformations of nodal Burniat surfaces with $K_S^2 = 4$ to extended Burniat surfaces with $K_S^2 = 4$ yield examples where $\mathrm{Def}(S,(\mathbb{Z}/2\mathbb{Z})^2) \to \mathrm{Def}(X,(\mathbb{Z}/2\mathbb{Z})^2)$ is not surjective.

Moreover, whereas for the canonical model we have:

$$Def(X, (\mathbb{Z}/2\mathbb{Z})^2) = Def(X),$$

for the minimal models we have

$$Def(S, (\mathbb{Z}/2\mathbb{Z})^2) \subsetneq Def(S)$$

and indeed the subset Def(S,(Z/2Z)2) corresponds to the locus of nodal Burniat surfaces.

The moduli space of pairs (S, G) of an extended (or nodal) smooth projective and minimal Burniat surface $S$with $K2S$=4, taken together with its canonical G:=(Z/2Z)2-action, is disconnected; but its image in the Gieseker moduli space is a connected open set.

10.5 Kuranishi subspaces for automorphisms of a fixed type

Proposition 174 is quite useful when one analyses the deformations of a given G-action, say on a compact complex manifold: it tells us that we have to look at the complex structures for which the given differentiable action is holomorphic. Hence we derive (see [86]):

Proposition 178 Consider a fixed action of a finite group G on a compact complex manifold X. Then we obtain a closed subset*Def*(X, G) of the Kuranishi space, corresponding to deformations which preserve the given action, and yielding a maximal family of deformations of the G-action.

The subset *Def*(X, G) is the intersection *Def*(X)∩$H1(\Theta X)G$.

Remark 179 *(1)*The proof is based on the well known and already cited Cartan's lemma ([78]), that the action of a finite group in an analytic neighbourhood of a fixed point can be linearized. The proof and the result extend also to encompass compact complex spaces.

(2)The above proposition is, as we shall now show in the example of projective curves, quite apt to estimate the dimension of the subspace *Def*(X,G)⊂*Def*(X). For instance, if *Def*(X) is smooth, then the dimension of *Def*(X, G) is just equal to $dim(H^1(\Theta X)^G)$.

We want to show how to calculate the tangent space to the Kuranishi space of G-invariant deformations in some cases.

To do this, assume that X is smooth, and that there is a normal crossing divisor $D=\cup iDi$ such that

(1)Di is a smooth divisor

(2)the Di 's intersect transversally

(3)there is a cyclic sugbroup $Gi\subset G$ such that $Di=Fix(Gi)$

(4)the stabilizer on the smooth locus of $Di1\cap Di2\cap\cdots\cap Dik$ equals $Gi1\oplus Gi2\oplus\cdots\oplus Gik$.

Then, setting $Y:=X/G$, Y is smooth and the branch divisor B is a normal crossing divisor $B=\cup iBi$.

Define as usual the sheaf of logarithmic forms $\Omega^1_X(\log D_1, \ldots, \log D_h)$, as the sheaf generated by Ω^1_X, and by the logarithmic derivatives of the equations si of the divisors Di, define $\Theta_Y(-\log D_1, \ldots, -\log D_h)$ as the dual sheaf (this sheaf can also be more generally defined as the sheaf of derivations carrying the ideal sheaf of D to itself). We have then the following proposition.

Proposition 180 Let $p:X\to Y=X/G$ the quotient projection. Then $p_*(\Theta_X)=\Theta_Y(-\log B_1,\dots,-\log B_h)$.

In particular, $H^i(\Theta_X)=H^i(\Theta_Y(-\log B_1,\dots,-\log B_h))$.

Idea of proof The basic idea of the proof is the calculation one does in dimension $n=1$ (and is the same one to be done at the generic point of the divisor D_i, which maps to B_i).

Namely, assume that the local quotient map is given by $w=z^m$, the action being given by $z\mapsto \epsilon z$, ϵ being a primitive m-th root of unity. It follows that $dz\mapsto \epsilon dz$, and, dually, $(\partial/\partial z)\mapsto \epsilon^{-1}(\partial/\partial z)$.

In particular,

$$d\log(z) = dz/z = (1/m)dw/w \quad and \quad z(\partial/\partial z) = mw(\partial/\partial w)$$

are invariant.

Then a vector field

$$\theta = f(z)\partial/\partial z = f(z)/z(z\partial/\partial z) = f(z)/zmw(\partial/\partial w)$$

is invariant if and only if $f(z)/z$ is invariant, hence if and only if $f(z)/z=F(w)$, where $F(w)$ is holomorphic. ∎

For instance, in the case of algebraic curves, where $Def(X)$ is smooth, the dimension of $Def(X, G)$ equals the dimension of

$$H^1(\Theta_Y(-\log B_1,\dots,-\log B_d)) = H^1(\Theta_Y(-y_1,\dots,-y_d)).$$

Let g' be the genus of Y, and observe that this vector space is Serre dual to $H^0(2K_Y+y_1+\dots+y_d))$. By Riemann-Roch its dimension equals $3g'-3+d$ whenever $3g'-3+d\geq 0$ (in fact $H^0(\Theta_Y(-y_1,\dots,-y_d))=0$ when $2-2g'-d<0$).

Calculations with the above sheaves are the appropriate ones to calculate the deformations of ramified coverings, see for instance [32, 82, 95, 317], and especially [33].

The main problem with the group of automorphisms on minimal models of surfaces (see [114]) is that a limit of isomorphisms need not be an isomorphism[12], it can be for instance a Dehn twist on a vanishing cycle (see also [106, 332]).

11 MODULI SPACES OF SYMMETRY MARKED VARIETIES

11.1 Moduli marked varieties

We give now the definition of a symmetry marked variety for projective varieties, but one can similarly give the same definition for complex or Kähler manifolds; to understand the concept of a marking, it suffices to consider a cyclic group acting on a variety X. A marking consists in this case of the choice of a generator for the group acting on X. The marking is very important when we have several actions of a group G on some projective varieties, and we want to consider the diagonal action of G on their product.

Definition 181 *(1)* A *G-marked (projective) variety* is a triple (X,G,η) where X is a projective variety, G is a group and $\eta:G\to \mathrm{Aut}(X)$ is an injective homomorphism;

(2)equivalently, a marked variety is a triple (X,G,α) where $\alpha:X\times G\to X$ is a faithful action of the group Gon X.

(3)Two marked varieties (X,G,α), (X',G,α') are said to be *isomorphic* if there is an isomorphism $f:X\to X'$transporting the action $\alpha:X\times G\to X$ into the action $\alpha':X'\times G\to X'$, i.e., such that

$$f\circ\alpha=\alpha'\circ(f\times \mathrm{id})\Leftrightarrow \eta'=Ad(f)\circ\eta,\quad Ad(f)(\phi):=f\phi f^{-1}.$$

(4)If G is a subset of $\mathrm{Aut}(X)$, then the natural marked variety is the triple (X, G, i), where $i:G\to\mathrm{Aut}(X)$ is the inclusion map, and it shall sometimes be denoted simply by the pair (X, G).

(5)A marked curve (D,G,η) consisting of a smooth projective curve of genus g and a faithful action of the group G on D is said to be a *marked triangle curve of genus* g if $D/G\cong$P1 and the quotient morphism $p:D\to D/G\cong$P1 is branched in three points.

Remark 182 Observe that:

(1)we have a natural action of $\mathrm{Aut}(G)$ on G-marked varieties, namely, if $\psi\in Aut(G)$,

$$\psi(X,G,\eta):=(X,G,\eta\circ\psi^{-1}).$$

The corresponding equivalence class of a G-marked variety is defined to be a *G-(unmarked) variety.*

(2)the action of the group Inn(G) of inner automorphisms does not change the isomorphism class of (X,G,η) since, for $\gamma \in G$, we may set $f:=\eta(\gamma)$, $\psi:=Ad(\gamma)$, and then $\eta \circ \psi = Ad(f) \circ \eta$, since $\eta(\psi(g)) = \eta(\gamma g \gamma^{-1}) = \eta(\gamma)\eta(g)(\eta(\gamma)^{-1}) = Ad(f)(\eta(g))$.

(3)In the case where G=Aut(X), we see that Out(G) acts simply transitively on the isomorphism classes of the Aut(G)-orbit of (X,G,η).

In the spirit of the above concept we have already given, in the previous section, the definition of a family of$G$-automorphisms (we shall also speak of a family of G-marked varieties).

The local deformation theory of a G-marked variety X, at least in the case where the group G is finite, is simply given by the fixed locus Def(X)G of the natural G -action on the Kuranishi space Def(X). As we mentioned previously (see 177), one encounters difficulties, when X has dimension at least 2, to use the Kuranishi approach for a global theory of moduli of G-marked minimal models.

But we have a moduli space of G-marked varieties in the case of curves of genus $g \geq 2$ and in the case of canonical models of surfaces of general type, and similarly also for canonical models in higher dimension.

In fact, one considers again a fixed linear representation space for the group G, the associated projective action,

$$\left(\mathbb{P}\left(V_m = \bigoplus_{\rho \in Irr(G)} W_\rho^{n(\rho)}\right), G\right).$$

and a locally closed subset of the Hilbert scheme

$$\mathcal{H}_{can}(\chi, K^2)^G \subset \mathcal{H}^0(\chi, K^2)$$
$$\mathcal{H}_{can}(\chi, K^2)^G := \{X | \omega_X \cong \mathcal{O}_X(1), \gamma(X) = X \quad \forall \gamma \in G,$$
$$X \quad is\ normal,\ with\ canonical\ singularities\ \}.$$

In this case one divides the above subset by the subgroup

$$\mathcal{C}(G) \subset GL(V_m)$$

which is the centralizer of $G \subset GL(Vm)$.

Remark 183 (1) Assume in fact that there is an isomorphism of the marked varieties (X, G) and (X',G) (here the marking is furnished by an m-th pluricanonical embedding of $X,X' \hookrightarrow P(Vm)$, and by the fixed action $\rho: G \hookrightarrow GL(Vm)$).

Then there is an isomorphism $f: X \cong X'$ with $f \circ \gamma = \gamma \circ f \ \forall \gamma \in G$. Now, f induces a linear map of Vm which we denote by the same symbol: hence we can write the previous condition as

$$f \circ \gamma \circ f^{-1} = \gamma, \ \forall \gamma \in G \Leftrightarrow f \in \mathcal{C}(G).$$

(2) By Schur's lemma $\mathcal{C}(G) = \Pi_{\rho \in Irr(G)} GL(n(\rho), \mathbb{C})$ is a reductive group.

(3) Observe that, since the dimension of Vm is determined by the holomorphic invariants of X, then there is only a finite number of possible representation types for the action of G on Vm.

(4) Assume moreover that X,X' are isomorphic varieties: then there is such a linear isomorphism $f: Vm \to Vm$ sending X to X'.

The variety X' has therefore two G-markings, the marking $\eta': G \to Aut(X')$ provided by the pluricanonical embeddding $X' \hookrightarrow P(Vm)$, and another one given, in an evident notation, by $Ad(f) \circ \eta$.

Therefore $Aut(X')$ contains two subgroups, $G := Im(\eta')$, and $G' := Im(Ad(f) \circ \eta)$. $G = G'$ if and only if f lies in the normalizer NG of G.

(5) This happens if G is a full subgroup (this means that $G = Aut(X')$). In general f will just belong to the normalizer NX' of $Aut(X')$, and NG shall just be the finite index subgroup of NX' which stabilises, via conjugation, the subgroup G.

(6) In other words, we define the marked moduli space (of a fixed representation type) as the quotient

$$\mathfrak{M}[G] := \mathcal{H}_{can}\left(\chi, K^2\right)^G / \mathcal{C}_G.$$

Then the forgetful map from the marked moduli space to the moduli space $\mathfrak{M}[G] \to \mathfrak{M}$, which factors through the quotient $\mathfrak{M}(G) \to \mathfrak{M}^G$, yields finite maps $\mathfrak{M}(G)$ by $\mathcal{N}_G / \mathcal{C}_G$.

We omit here technical details in higher dimension, which have been treated by Binru Li in his Bayreuth Ph.D. Thesis (in preparation).

Observe that the action of *Out* (G) does not need to respect the representation type.

Let us see now how the picture works in the case of curves: this case is already very enlightening and intriguing.

11.2 Moduli of curves with automorphisms

There are several 'moduli spaces' of curves with automorphisms. First of all, given a finite group G, we define a subset $\mathfrak{M}_{g,G}$ of the moduli space Mg of smooth curves of genus $g>1$: $\mathfrak{M}_{g,G}$ is the locus of the curves that admit an effective action by the group G. It turns out that Mg,G is a Zariski closed algebraic subset.

In order to understand the irreducible components of Mg,G we have seen that Teichmüller theory plays an important role: it shows the connectedness, given an injective homomorphism ρ:G→$Mapg$, of the locus

$$\mathcal{T}_{g,\rho} := Fix(\rho(G)).$$

Its image Mg,ρ in Mg,G is a Zariski closed irreducible subset (as observed in [115]). Recall that to a curve $C$of genus g with an action by G we can associate several discrete invariants that are constant under deformation.

The first is the above *topological type* of the G-action: it is a homomorphism ρ:G→$Mapg$, which is well-defined up to inner conjugation (induced by different choices of an isomorphism $Map(C)$≅$Mapg$).

We immediately see that the locus Mg,ρ is first of all determined by the subgroup $\rho(G)$ and not by the marking. Moreover, this locus remains the same not only if we change ρ modulo the action by Aut(G), but also if we change ρ by the adjoint action by Map$g$.

Definition 184 (1) The moduli space of G-marked curves of a certain topological type ρ is the quotient of the Teichmüller submanifold Tg,ρ by the centralizer subgroup Cρ(G) of the subgroup ρ(G) of the mapping class group. We get a normal complex space which we shall denote $\mathfrak{M}_g[\rho]$. $\mathfrak{M}_g[\rho] = \mathcal{T}_{g,\rho}/C_{\rho(G)}$ is a finite covering of a Zariski closed subset of the usual moduli space (its image Mg,ρ), therefore it is quasi-projective, by the theorem of Grauert and Remmert.

(2)Defining Mg(ρ) as the quotient of Tg,ρ by the normalizer Nρ(G) of ρ(G), we call it the moduli space of curves with a G-action of a given topological type. It is again a normal quasi-projective variety.

Remark 185(1)If we consider G′:=ρ(G) as a subgroup G′⊂Mapg, then we get a natural G′-marking for any C∈Fix(G′)=Tg,ρ.

(2)As we said, *Fix*(G')=Tg,ρ is independent of the chosen marking, moreover the projection $Fix(G') = \mathcal{T}_{g,\rho} \rightarrow \mathfrak{M}_{g,\rho}$ factors through a finite map Mg(ρ)→Mg,ρ.

The next question is whether Mg(ρ) maps 1-1 into the moduli space of curves. This is not the case, as we shall easily see. Hence we give the following definition.

Definition 186 Let *G⊂Mapg* be a finite group, and let *C* represent a point in *Fix*(*G*). Then we have a natural inclusion *G*→*AC*:=*Aut*(*C*), and *C* is a fixed point for the subgroup *AC⊂Mapg*: *AC* is indeed the stabilizer of the point*C* in *Mapg*, so that locally (at the point of Mg corresponding to *C*) we get a complex analytic isomorphism Mg=Tg/*AC*.

We define *HG*:=∩*C*∈*Fix*(*G*)*AC* and we shall say that *G* is a **full subgroup** if *G*=*HG*. Equivalently, *HG* is the largest subgroup *H* such that *Fix*(*H*)=*Fix*(*G*).

This implies that *HG* is a full subgroup.

Remark 187 The above definition shows that the map $T_{g,\rho} \to \mathfrak{M}_{g,\rho}$ factors through Mg(HG):=Tg,ρ/NHG, hence we restrict our attention only to full subgroups.

Proposition 188 If *H* is a full subgroup *H*⊂*Mapg*, and *ρ*:*H*⊂*Mapg* is the inclusion homomorphism, then M*g*(*ρ*) is the normalization of M*g*,*ρ*.

Proof Recall that $\mathfrak{M}_g(H) = Fix(H)/\mathcal{N}_H$, and that for a general curve *C*∈*Fix*(*H*) we have *AC*=*H*.

Assume then that *C*=*γ*(*C*′), and that both *C*,*C*′∈*Fix*(*H*). Then *γACγ*−1=*AC*, hence for *C* general *γ*∈N*H*.

This means that the finite surjective morphism $\varphi : \mathfrak{M}_g(H) \to \mathfrak{M}_{g,}$*H* is generically bijective; M*g*(*H*) is a normal variety, being locally the quotient of a smooth variety by a finite group: hence *φ* is the normalization morphism and $\mathfrak{M}_g(H)$is the normalization of M*g*,*H*. ▯

Next we investigate when the natural morphism M*g*(*H*)→M*g*,*H* is not injective. In order that this be the case, we have already seen that we must pick up a curve *C* such that *H*⊂*AC*≠*H*. Now, let *C*′ be another point of*Fix* (*H*) which has the same image in the moduli space: this means that *C*′ is in the orbit of *C* so there is an element *γ*∈*Mapg* carrying *C*′ to *C*.

By conjugation *γ* sends *H* to another subgroup *H*′ of *AC*. We can assume that *H*≠*H*′, else *γ* lies in the normalizer *NH* and we have the same point in M*g*(*H*). Hence the points of M*g*(*H*) that have the same image as *C* correspond to the subset of subgroups *H*′⊂*AC* which are conjugate to *H* by the action of some *γ*∈*Mapg*.

Example 189 Consider the genus 2 curve *C* birational to the affine curve with equation *y*2=(*x*6−1). Its canonical ring is generated by *x*0,*x*1,*y* and is the quotient $\mathbb{C}[x_0, x_1, y]/(y^2 - x_1^6 + x_0^6)$.

Its group of automorphisms is generated (as a group of projective transformations) by *a*(*x*0,*x*1,*y*):=(*x*0,*ϵx*1,*y*)where *ϵ* is a primitive sixth root of 1, and by *b*(*x*0,*x*1,*y*):=(*x*1,*x*0,*iy*). *a* has order 6, *b* has order 4, the square *b*2 is the hyperelliptic involution *h*(*x*0,*x*1,*y*):=(*x*0,*x*1,−*y*).

We have

$$b \circ a \circ b^{-1} = a^{-1},$$

a formula which implies the (known) fact that the hyperelliptic involution lies in the centre of the group $AC=Aut(C)$.

Taking as H the cyclic subgroup generated by b, the space $Fix\ (H)$ has dimension equal to 1 since $C \to C/H=$P1 is branched in 4 points. The quotient of $AC=Aut(C)$ by $b2$ is the dihedral group $D6$, and we see that H is conjugate to 6 different subgroups of AC.

Of course an important question in order to understand the locus in Mg of curves with automorphisms is the determination of all the non full subgroups $G \neq HG$.

For instance Cornalba [127] answered this question for cyclic groups of prime order, thereby obtaining a full determination of the irreducible components of $Sing$(Mg). In fact, for $g \geq 4$, the locus $Sing$(Mg) is the locus of curves admitting a nontrivial automorphism, so this locus is the locus of curves admitting a nontrivial automorphism of prime order. In this case, as we shall see in the next subsection, the topological type is easily determined. We omit to state Cornalba's result in detail: it amounts in fact to a list of all non full such subgroups of prime order. We limit ourselves to indicate the simplest example of such a situation.

Example 190 Consider the genus $g = \frac{p-1}{2}$ curve C, birational to the affine curve with equation $zp=(x2-1)$, where p is an odd prime number. Letting H be the cyclic group generated by the automorphisms $(x,z) \mapsto (x,\epsilon z)$, where ϵ is a primitive p-th root of unity, we see that the quotient morphism $C \to C/H \cong$P1 corresponds to the inclusion of fields C(x)$\hookrightarrow$C(C). The quotient map is branched on the three points $x=1,-1,\infty$, so that $Fix\ (H)$ consists of just a point (the above curve C).

An easy inspection of the above equation shows that the curve C is hyperelliptic, with involution $h:(x,z) \mapsto (-x,z)$ which is hyperelliptic since C(C)h=C(z). In this case the locus $Fix(H)$ is contained in the hyperelliptic locus, and therefore it is not an irreducible component of $Sing$(Mg).

11.3 Numerical and homological invariants of group actions on curves

As already mentioned, given an effective action of a finite group G on C, we set $C' := C/G$, $g' := g(C')$, and we have the quotient morphism $p: C \to C/G =: C'$, a G-cover.

The geometry of p encodes several numerical invariants that are constant on $M_{g,\rho}(G)$: first of all the genus g' of C', then the number d of branch points $y_1,\dots,y_d \in C'$.

We call the set $B=\{y_1,\dots,y_d\}$ the branch locus, and for each y_i we denote by m_i the multiplicity of y_i (the greatest number dividing the divisor $p^*(y_i)$). We choose an ordering of B such that $m_1 \leq \dots \leq m_d$.

These numerical invariants $g', d, m_1 \leq \dots \leq m_d$ form the so-called **primary numerical type**.

$p: C \to C'$ is determined (Riemann's existence theorem) by the monodromy, a surjective homomorphism:

$$\mu: \pi_1(C' \setminus B) \to G.$$

We have:

$$\pi_1(C' \setminus B) \cong \Pi_{g',d} := \langle \gamma_1, \dots, \gamma_d, \alpha_1, \beta_1, \dots, \alpha_{g'}, \beta_{g'} | \prod_{i=1}^{d} \gamma_i \prod_{j=1}^{g'} [\alpha_j, \beta_j] = 1 \rangle.$$

We set then $c_i := \mu(\gamma_i)$, $a_j := \mu(\alpha_j)$, $b_j := \mu(\beta_j)$, thus obtaining a Hurwitz generating vector, i.e. a vector

$v := (c_1,\dots,c_d,a_1,b_1,\dots,a_{g'},b_{g'}) \in G^{d+2g'}$

s.t.

- G is generated by the entries $c_1,\dots,c_d,a_1,b_1,\dots,a_{g'},b_{g'}$,
- $c_i \neq 1_G$, $\forall i$, and
- $\prod_{i=1}^{d} c_i \prod_{j=1}^{g'} [a_j, b_j] = 1.$

We see that the monodromy μ is completely equivalent, once an isomorphism $\pi_1(C' \setminus B) \cong \Pi_{g',d}$ is chosen, to the datum of a Hurwitz generating vector (we also call the sequence $c_1,\dots,c_d,a_1,b_1,\dots,a_{g'},b_{g'}$ of the vector's coordinates a *Hurwitz generating system*).

A second numerical invariant of these components of $\mathfrak{M}_g(G)$ is obtained from the monodromy $\mu: \pi_1(C' \setminus \{y_1,\dots,y_d\}) \to G$ of the restriction of p to $p^{-1}(C' \setminus \{y_1,\dots,y_d\})$, and is called the *ν-type* or Nielsen function of the covering.

The Nielsen function ν is a function defined on the set of conjugacy classes in G which, for each conjugacy class C in G, counts the number ν(C) of local monodromies $c1,\ldots,cd$ which belong to C (observe that the numbers $m1\leq\cdots\leq md$ are just the orders of the local monodromies).

Observe in fact that the generators γj are well defined only up to conjugation in the group $\pi 1(C'\backslash\{y1,\ldots,yd\})$, hence the local monodromies are well defined only up to conjugation in the group G.

We have already observed that the irreducible closed algebraic sets $Mg,\rho(G)$ depend only upon what we call the 'unmarked topological type', which is defined as the conjugacy class of the subgroup $\rho(G)$ inside $Mapg$. This concept remains however still mysterious, due to the complicated nature of the group $Mapg$. Therefore one tries to use more geometry to get a grasp on the topological type.

The following is immediate by Riemann's existence theorem and the irreducibility of the moduli space $\mathfrak{M}_{g',d}$ of d-pointed curves of genus g'. Given g' and d, the unmarked topological types whose primary numerical type is of the form $g',d,m1,\ldots,md$ are in bijection with the quotient of the set of the corresponding monodromies μ modulo the actions by $Aut(G)$ and by $Map(g',d)$.

Here $Map(g',d)$ is the full mapping class group of genus g' and d unordered points.

Thus Riemann's existence theorem shows that the components of the moduli space

$$\mathfrak{M}(G) := \cup_g \mathfrak{M}_g(G)$$

with numerical invariants g',d correspond to the following quotient set.

Definition 191

$$\mathcal{A}(g', d, G) := \mathrm{Epi}(\Pi_{g',d}, G)/\mathrm{Map}_{g',d} \times \mathrm{Aut}(G).$$

Thus a first step toward the general problem consists in finding a fine invariant that distinguishes these orbits.

In the paper [115] we introduced a new homological invariant $\hat{\epsilon}$ for G-actions on smooth curves (and showed that, in the case where G is the dihedral group Dn of order $2n$, $\hat{\epsilon}$ is a fine invariant since it distinguishes the different unmarked topological types).

This invariant generalizes the classical homological invariant in the unramified case.

Definition 192 Let $p:C\to C/G=:C'$ be unramified, so that $d=0$ and we have a monodromy $\mu:\pi 1(C')\to G$.

Since C' is a classifying space for the group $\pi g'$, we obtain a continuous map

$$m : C' \to BG,\ \pi_1(m) = \mu.$$

Moreover, $H2(C',\mathbb{Z})$ has a natural generator $[C']$, the fundamental class of C' determined by the orientation induced by the complex structure of C'.

The homological invariant of the G-marked action is then defined as:

$$\epsilon := H_2(m)([C']) \in H_2(BG, \mathbb{Z}) = H_2(G, \mathbb{Z}).$$

If we forget the marking we have to take ϵ as an element in $H2(G,\mathbb{Z})/Aut(G)$.

Proposition 193 Assume that a finite group G has a fixed point free action on a curve of genus $g\geq 3$.

Let $p:C\to C/G=:C'$ be the quotient map and pick an isomorphism $\pi 1(C')\cong \pi g'$. Let $\mu:\pi g'\to G$ be the surjection corresponding to the monodromy of p, and denote by $ai:=\mu(\alpha i),bj:=\mu(\beta j)$, $i,j=1,\dots,g'$.

Assume that we have a realization $G=F/R$ of the group G as the quotient of a free group F, and that $\hat{a}i,\hat{b}j$are lifts of ai,bj to F.

Then the homological invariant ϵ of the covering is the image of

$$\epsilon' = \prod_1^{g'} [\widehat{a_i}, \widehat{b_i}] \in [F, F] \cap R$$

into $H2(G,\mathbb{Z})=([F,F]\cap R)/[F,R]$.

Proof C' is obtained attaching a 2-cell D to a bouquet of circles, with boundary ∂D mapping to $\prod_1^{g'} [\alpha_i, \beta_i]$.

Similarly the 2-skeleton BG^2 of BG is obtained from a bouquet of circles, with fundamental group $\cong F$, attaching 2-cells according to the relations in R. Since $\prod_1^{g'} [a_i, b_i] = 1$ in $\pi g'$, the relation $\prod_1^{g'} [\widehat{a_i}, \widehat{b_i}]$ is a product of elements of R. In this way m is defined on D, and the image of the fundamental class of C' is the image of D, which is a sum of 2-cells whose boundary is exactly the loop ϵ'. If we write ϵ as a sum of 2-cells, we get an element in $H2(BG2,BG1,Z)$: but since ϵ is a product of commutators, the boundary of the corresponding 2-chain is indeed zero, so we get an element in $H2(BG,Z)=H2(G,Z)$.

11.4 The refined homology invariant in the ramified case

Assume now that $p:C\to C/G$ is ramified. Then we define H to be the minimal normal subgroup of Ggenerated by the local monodromies $c1,\dots,cd \Leftrightarrow H$ is the (normal) subgroup generated by the transformations $\gamma\in G$ which have some fixed point.

We have therefore a factorization of p

$C\to C'':=C/H\to C':=C/G$

where $C''\to C'$ is an unramified G / H-cover, and where $C\to C''$ is totally ramified.

The refined homology invariant includes and extends two invariants that have been studied in the literature, and were already mentioned: the ν-type (or Nielsen type) of the cover (also called shape in [170]) and the class in the second homology group $H2(G/H,Z)$ corresponding to the unramified cover $p':C''=C/H\to C'$.

The construction of the invariant $\hat{}\epsilon$ is similar in spirit to the procedure used in the unramified case. But we achieve a little less in the 'branched' case of a non-free action. In this case we are only able to associate, to two given actions with the same ν-type, an invariant in a quotient group of $H2(G,Z)$ which is the 'difference' of the respective $\hat{}\epsilon$-invariants. Here is the way we do it.

Let Σ be the Riemann surface (with boundary) obtained from C' after removing open discs Δi around each of the branch points.

Take generators $\gamma1,\dots,\gamma d$ formed by a simple path going from the base point $y0$ to a point zi on the circle bounding the open discs Δi, and by the circle $\partial\Delta i$.

Fix then a CW-decomposition of Σ as follows. The 0-skeleton $\Sigma 0$ consists of the point $y0$ and of the points zi, i=1,...,d. The 1-skeleton $\Sigma 1$ is given by the geometric basis $\alpha 1,\ldots,\beta g',\gamma 1,\ldots,\gamma d$ and the 2-skeleton $\Sigma 2$consists of one cell.

The restriction $p\Sigma$ of $p{:}C\to C'$ to $p-1(\Sigma)$ is an unramified G-covering of Σ and hence corresponds to a continuous map $Bp\Sigma{:}\Sigma\to BG$, well defined up to homotopy. Let $Bp1{:}\Sigma\to BG$ be a cellular approximation of $Bp\Sigma$. Since $Bp1$ can be regarded as a map of pairs $Bp1{:}(\Sigma,\partial\Sigma)\to(BG,BG1)$, the push-forward of the fundamental (orientation) class $[\Sigma,\partial\Sigma]$ gives an element

$$Bp_{1*}[\Sigma,\partial\Sigma]\in H_2(BG,BG^1)=\frac{R}{[F,R]}.$$

This element depends on the chosen cellular approximation $Bp1$ of $Bp\Sigma$, but its image in a quotient group which we denote by $K\Gamma$ does not depend on the chosen cellular approximation, as shown in [116]. The main idea to define this quotient group is that a homotopy between two 1-cellular approximations can also be made cellular, hence the difference in relative homology is the sum of boundaries of cylinders; and if we have a cylinder C with upper circle a, lower circle c (with the same orientation), and meridian b, then the boundary of the corresponding 2-cell is

$$\partial C=abc^{-1}b^{-1}.$$

Definition 194 (1) For any finite group G, let F be the free group generated by the elements of G and let $R\trianglelefteq F$ be the subgroup of relations, that is $G=F/R$.

Denote by $\hat{}g\in F$ the generator corresponding to $g\in G$.

For any $\Gamma\subset G$, union of non trivial conjugacy classes, let $G\Gamma$ be the quotient group of F by the minimal normal subgroup $R\Gamma$ generated by $[F, R]$ and by the elements $\hat{a}\hat{b}\hat{c}^{-1}\hat{b}^{-1}\in F$, for any $a,c\in\Gamma$, $b\in G$, such that $b-1ab=c$.

(2) Define instead

$$K_\Gamma=R/R_\Gamma\subset G_\Gamma=F/R_\Gamma,$$

a central subgroup of $G\Gamma$. We have thus a central extension

$$1 \to K_\Gamma \to G_\Gamma \to G \to 1.$$

(3) Define

$$H_{2,\Gamma}(G) = \ker\left(G_\Gamma \to G \times G_\Gamma^{ab}\right).$$

Notice that

$$H_2(G, \mathbb{Z}) \cong \frac{R \cap [F, F]}{[F, R]} \cong \ker\left(\frac{F}{[F, R]} \to G \times G_\emptyset^{ab}\right).$$

Remark 195 In particular, when Γ=∅, *H*2,Γ(*G*)≅*H*2(*G*,Z).

By [115, Lemma 3.12] we have that the morphism

$$R \cap [F, F] \to \frac{R}{R_\Gamma}, \quad r \mapsto rR_\Gamma$$

induces a surjective group homomorphism

$$H_2(G, \mathbb{Z}) \to H_{2,\Gamma}(G).$$

Now, to a given *G*-cover *p*:*C*→*C*′ we associate the set Γ of the local monodromies, i.e., of the elements of*G* which can be geometrically described as those which (i) stabilize some point *x* of *C* and (ii) act on the tangent space at *x* by a rotation of angle $\frac{2\pi}{m}$ where *m* is the order of the stabilizer at *x*.

In terms of the notation that we have previously introduced Γ=Γ*v* is simply the union of the conjugacy classes of the *ci*'s.

Definition 196 The tautological lift ˆ*v* of *v* is the vector:

$$(\widehat{c_1}, \ldots, \widehat{c_d}; \widehat{a_1}, \widehat{b_1}, \ldots, \widehat{a_{g'}}, \widehat{b_{g'}}).$$

Finally, define ϵ(*v*) as the class in *K*Γ of

$$\prod_1^d \widehat{c_j} \cdot \prod_1^{g'} [\widehat{a_i}, \widehat{b_i}].$$

It turns out that the image of $\epsilon(v)$ in $K\Gamma/Inn(G)$ is invariant under the action of $Map(g',d)$, as shown in [115].

Moreover the ν-type of a Hurwitz monodromy vector v can be deduced from $\epsilon(v)$, as it is essentially the image of $\epsilon(v)$ in the abelianized group $Gab\Gamma$.

Proposition 197 Let v be a Hurwitz generating vector and let $\Gamma v \subset G$ be the union of the conjugacy classes of the cj's , $j \leq d$. The abelianization $G^{ab}_{\Gamma_v}$ *of* G_{Γ_v} can be described as follows:

$$G^{ab}_{\Gamma_v} \cong \left(\bigoplus_{C \subset \Gamma_v} \mathbb{Z}\langle C \rangle \right) \oplus \left(\bigoplus_{g \in G \setminus \Gamma_v} \mathbb{Z}\langle g \rangle \right),$$

where C denotes a conjugacy class of G.

Moreover the Nielsen function $\nu(v)$ coincides with the vector whose C-components are the corresponding components of the image of $\epsilon(v) \in G_{\Gamma_v}$ *in* $G^{ab}_{\Gamma_v}$.

11.5 Genus stabilization of components of moduli spaces of curves with *G*-symmetry

In order to take into account also the automorphism group $Aut(G)$, one has to consider

$$K^{\cup} := \coprod_{\Gamma} K_{\Gamma},$$

the disjoint union of all the $K\Gamma$'s. Now, the group $Aut(G)$ acts on $K\cup$ and we get a map

$$\hat{\epsilon} : \mathcal{A}(g', d, G) \to (K^{\cup}) / _{Aut(G)}$$

which is induced by $v \mapsto \epsilon(v)$.

Next, one has to observe that the Nielsen functions of coverings have to satisfy a necessary condition, consequence of the relation

$$\prod_{i=1}^{d} c_i \prod_{j=1}^{g'} [a_j, b_j] = 1.$$

Definition 198 **An element**

$$\nu = (n_C)_C \in \bigoplus_{C \neq \{1\}} \mathbb{N}\langle C \rangle$$

is **admissible** if the following equality holds in the Z-module *Gab*:

$$\sum_C n_C \cdot [C] = 0$$

(here [C] denotes the image element of C in the abelianization *Gab*).

The main result of [116] is the following 'genus stabilization' theorem.

Theorem 199 **There is an integer *h* such that for *g*′>*h***

$$\hat{\epsilon} : \mathcal{A}(g', d, G) \to (K^{\cup})/_{Aut(G)}$$

induces a bijection onto the set of admissible classes of refined homology invariants.

In particular, if $g'>h$, and we have two Hurwitz generating systems $v1$,$v2$ having the same Nielsen function, they are equivalent if and only if the 'difference' ˆϵ($v1$)ˆϵ($v2$)−1∈$H2$,Γ(G) is trivial.

The above result extends a nice theorem of Livingston [272], Dunfield and Thurston [141] in the unramifed case, where also the statement is simpler.

Theorem 200 **For *g*′>>0**

$$\hat{\epsilon} : \mathcal{A}(g', 0, G) \to H_2(G, \mathbb{Z})/_{Aut(G)}$$

is a bijection.

Remark 201 Unfortunately the integer *h* in Theorem 199, which depends on the group *G*, is not explicit.

A key concept used in the proof is the concept of genus stabilization of a covering, which we now briefly explain.

Definition 202 Consider a group action of G on a projective curve C, and let $C \to C' = C/G$ the quotient morphism, with monodromy

$$\mu : \pi_1(C' \backslash B) \to G$$

(here B is as usual the branch locus). Then the first genus stabilization of the differentiable covering is defined geometrically by simply adding a handle to the curve C', on which the covering is trivial.

Algebraically, given the monodromy homomorphism

$$\mu : \pi_1(C' \backslash B) \cong \Pi_{g',d} := \langle \gamma_1, \dots, \gamma_d, \alpha_1, \beta_1, \dots, \alpha_{g'}, \beta_{g'} | \times \prod_{i=1}^{d} \gamma_i \prod_{j=1}^{g'} [\alpha_j, \beta_j] = 1 \rangle \to G,$$

we simply extend μ to $\mu^1 : \Pi_{g'+1,d} \to G$ setting

$$\mu^1(\alpha_{g'+1}) = \mu^1(\beta_{g'+1}) = 1_G.$$

In terms of Hurwitz vectors and Hurwitz generating systems, we replace the vector

$$v := (c_1, \dots, c_d, a_1, b_1, \dots, a_{g'}, b_{g'}) \in G^{d+2g'}$$

by

$$v^1 := (c_1, \dots, c_d, a_1, b_1, \dots, a_{g'}, b_{g'}, 1, 1) \in G^{d+2g'+2}.$$

The operation of first genus stabilization generates then an equivalence relation among monodromies (equivalently, Hurwitz generating systems), called **stable equivalence**.

The most important step in the proof, the geometric understanding of the invariant $\epsilon \in H2(G,Z)$ was obtained by Livingston [272].

Theorem 203 Two monodromies $\mu 1$, $\mu 2$ are stably equivalent if and only if they have the same invariant $\epsilon \in H2(G,Z)$.

A purely algebraic proof of Livingston's theorem was given by Zimmermann [380], while a nice sketch of proof was given by Dunfield and Thurston [141].

Idea of proof Since one direction is clear (the map to BG is homotopically trivial on the handle that one adds to C'), one has to show that two coverings are stably equivalent if their invariant is the same.

The first idea is then to interpret second homology as bordism: given two maps of two curves $C'1, C'2 \to BG$, they have the same invariant in $H2(G,Z)$ (image of the fundamental classes $[C'1],[C'2]$) if and only if there is a 3-manifold W with boundary $\partial W = C'1 - C'2$, and a continuous map $f: W \to BG$ which extends the two maps defined on the boundary $\partial W = C'1 - C'2$.

Assume now that there is relative Morse function for $(W, \partial W)$, and that one first adds all the 1-handles (for the critical points of negativity 1) and then all the 2-handles.

Assume that at level t we have a curve $C't$, to which we add a 1-handle. Then the genus of $C't$ grows by 1, and the monodromy is trivial on the meridian α (image of the generator of the fundamental group of the cylinder we are attaching).

Pick now a simple loop β' meeting α transversally in one point with intersection number 1: since the monodromy μ is surjective, there is a simple loop γ disjoint from α and β' with $\mu(\gamma) = \mu(\beta')$. Replacing β' by $\beta := \beta'\gamma - 1$, we obtain that the monodromy is trivial on the two simple loops α and β. A suitable neighbourhood of $\alpha \cup \beta$ is then a handle that is added to $C't$, and with trivial monodromy. This shows that at each critical value of the Morse function we pass from one monodromy to a stably equivalent one (for the case of a 2-handle, repeat the same argument replacing the Morse function F by its opposite $-F$).

11.6 Classification results for certain concrete groups

The first result in this direction was obtained by Nielsen [313] who proved that ν determines ρ if G is cyclic (in fact in this case $H2(G,Z)=0$!).

In the cyclic case the Nielsen function for $G = Z/n$ is simply a function $\nu: (Z/n) \setminus \{0\} \to N$, and admissibility here simply means that

$$\sum_i i \cdot \nu(i) \equiv 0 \ (mod\ n).$$

The class of v is just the equivalence class for the equivalence relation $v(i) \sim vr(i), \forall r \in (\mathbb{Z}/n)^*$, where $vr(i) := v(ri), \forall i \in (\mathbb{Z}/n)$.

From the refined Nielsen realization theorem of [96] (37) it follows that the components of $\mathfrak{M}_g(\mathbb{Z}/n)$ are in bijection with the classes of Nielsen functions (see also [113] for an elementary proof).

Example 204 For instance, in the case $n=3$, the components of $\mathfrak{M}_g(\mathbb{Z}/3)$ correspond to triples of integers $g', a, b \in \mathbb{N}$ such that $a \equiv b \ (mod\ 3)$, and $g-1 = 3(g'-1)+a+b$ $(a := v(1), b := v(2))$.

For $g'=0$ we have, if we assume $a \leq b$ (we can do this by changing the generator of $\mathbb{Z}/3$), $b = a+3r, r \in \mathbb{N}$, and $2a+3r = g+2$, thus $a \equiv -(g-1) \ (mod\ 3)$, and $2a \leq g+2$.

In the case where $g'=0$, two such components N_a (labelled by a as above) have images which do not intersect in $\mathfrak{M}_g$ as soon as $g \geq 5$.

Otherwise we would have two $\mathbb{Z}/3$-quotient morphisms $p_1, p_2 : C \to \mathbb{P}^1$ and a birational map $C \to C'' \subset \mathbb{P}^1 \times \mathbb{P}^1$. Then C would be the normalization of a curve of arithmetic genus 4, so $g \leq 4$.

The genus g' and the Nielsen class (which refine the primary numerical type), and the homological invariant $h \in H_2(G/H, \mathbb{Z})$ (here H is again the subgroup generated by the local monodromies) determine the connected components of $\mathfrak{M}_g(G)$ under some restrictions: for instance when G is abelian or when G acts freely and is the semi-direct product of two finite cyclic groups (as it follows by combining results from [96,113, 145, 146]).

Theorem 205 (Edmonds) v and $h \in H_2(G/H, \mathbb{Z})$ determine ρ for G abelian. Moreover, if G is split-metacyclic and the action is free, then h determines ρ.

However, in general, these invariants are not enough to distinguish unmarked topological types, as one can see already for non-free D_n-actions (see [115]). Already for dihedral groups, one needs the refined homological invariant $\hat{\epsilon}$.

Theorem 206 ([115]) For the dihedral group $G=Dn$ the connected components of the moduli space $\mathfrak{M}_g$ (Dn) are in bijection, via the map ˆϵ, with the admissible classes of refined homology invariants.

Remark 207 In this case, the classification is simple: two monodromies with the same Nielsen function differ by an element in $H2,\Gamma(Dn)$, which is a quotient of $H2(Dn,\mathrm{Z})$. This last group is 0 if n is odd, and $\cong\mathrm{Z}/2$ for n even.

More precisely: $H2,\Gamma(Dn)=\{0\}$ if and only if

- n is odd or
- n is even and Γ contains some reflection or
- n is even and Γ contains the non-trivial central element.

In the remaining cases, $H2,\Gamma(Dn)=\mathrm{Z}/2\mathrm{Z}$.

The above result completes the classification of the unmarked topological types for $G=Dn$, begun in [112]; moreover this result entails the classification of the irreducible components of the loci Mg,Dn (see the appendix to [115]).

It is an interesting question: for which groups G does the refined homology invariant ˆϵ determine the connected components of $\mathfrak{M}_g(G)$?

In view of Edmonds' result in the unramified case, it is reasonable to expect a positive answer for split metacyclic groups (work in progress by Sascha Weigl) or for some more general metacyclic or metaabelian groups.

As mentioned in [141], p. 499, the group $G=\mathrm{P}SL(2,\mathrm{F}13)$ shows that, for $g'=2$, in the unramified case there are different components with trivial homology invariant $\epsilon\in H2(G,\mathrm{Z})$: these topological types of coverings are therefore stably equivalent but not equivalent.

11.7 *Sing*(M*g*) II: loci of curves with automorphisms in $\mathfrak{M}_g$

Several authors independently found restrictions in order that a finite subgroup H of *Mapg* be a not full subgroup (Singerman [343], Ries [327], and Magaard, Shaska, Shpectorov, and Völklein [275])

We refer to lemma 4.1 of [275] for the proof of the following result.

Theorem 208 (MSSV) Suppose *H*⊂*G*⊂*Mapg* and assume *Z*:=*Fix*(*H*)=*Fix*(*G*)⊂T*g*, with *H* a proper subgroup of *G*, and let *C*∈*Z*. Then

$$\delta := dim(Z) \leq 3.$$

(I) if δ=3, then *H* has index 2 in *G*, and *C*→*C*/*G* is covering of P1 branched on six points, *P*1,...,*P*6, and with branching indices all equal to 2. Moreover the subgroup *H* corresponds to the unique genus two double cover of P1 branched on the six points, *P*1,...,*P*6 (by Galois theory, intermediate covers correspond to subgroups of *G* bijectively).

(II) If δ=2, then *H* has index 2 in *G*, and *C*→*C*/*G* is covering of P1 branched on five points, P^1,...,P^5, and with branching indices 2,2,2,2,*c*5 (where obviously *c*5≥2). Moreover the subgroup *H* corresponds to a genus one double cover of P1 branched on four of the points *P*1,...,*P*5 which have branching index 2.

(III) If δ=1, then there are three possibilities.

(III-a) *H* has index 2 in *G*, and *C*→*C*/*G* is covering of P^1 branched on four points P^1,...,P^4, with branching indices 2,2,2,2*d*4, where *d*4>1. Moreover the subgroup *H* corresponds to the unique genus one double cover of P1 branched on the four points, P^1,...,P^4.

(III-b) *H* has index 2 in *G*, and *C*→*C*/*G* is covering of P1 branched on four points, P^1,...,P^4, with branching indices 2,2,*c*3,*c*4, where 2≤*c*3≤*c*4>2. Moreover the subgroup *H* corresponds to a genus zero double cover of P1 branched on two points whose branching index equals 2.

(III-c) *H* is normal in *G*, *G*/*H*≅(Z/2)2, moreover *C*→*C*/*G* is covering of P1 branched on four points, *P*1,...,*P*4, with branching indices 2,2,2,*c*4, where *c*4>2. Moreover the subgroup *H* corresponds to the unique genus zero cover of P1 with group (Z/2)2 branched on the three points *P*1,*P*2,*P*3 whose branching index equals 2.

The main point in the theorem above is a calculation via the Hurwitz formula, showing that the normalizer of*H* in *G* is nontrivial; from this follows the classification of the singular locus of M*g*, due to Cornalba, and which we do not reproduce here (see [127], and also [113] for a slightly different proof).

11.8 Stable curves and their automorphisms, *Sing* $(\overline{\mathfrak{M}_g})$

The compactification of the moduli space of curves M*g* is given by the moduli space ¯M*g* of stable curves of genus *g* (see [136, 179, 304]).

Definition 209 A stable curve of genus *g*≥2 is a reduced and connected projective curve *C*, not necessarily irreducible, whose singularities are only nodes, such that

(1)the dualizing sheaf *ωC* has degree 2*g*−2

(2)its group of automorphisms *Aut*(*C*) is finite (equivalently, each smooth irreducible component *D*⊂*C* of genus zero intersects *C**D* in at least 3 points).

Again the Kuranishi family *Def*(*C*) of a stable curve is smooth of dimension 3*g*−3, and the locus of stable curves which admit a given *G*-action forms a smooth submanifold of *Def*(*C*).

Hence again the singular locus of $\overline{\mathfrak{M}_g}$ corresponds to loci of curves with automorphisms which do not form a divisor.

Example 210 We say that *C* has an elliptic tail if we can write *C*=*C*1∪*E*, where *E* is a smooth curve of genus 1 intersecting *C*1 in only one point *P*.

In this case *C* has always an automorphism *γ* of order 2: *γ* is the identity on *C*1, and, if we choose *P* as the origin of the elliptic curve (*E*, *P*), *γ* is simply multiplication by −1.

An easy calculation shows that such curves with an elliptic tail form a divisor inside $\overline{\mathfrak{M}_g}$.

We refer to [113] (the second part) for the explicit determination of the loci of stable curves admitting an action by a cyclic group of prime order, especially those contained in the boundary $\partial\overline{\mathfrak{M}_g}$. From this follows the description of the singular locus *Sing*$\partial\overline{\mathfrak{M}_g}$, which we do not reproduce here.

An interesting question is to describe similar loci also for other groups.

As mentioned, there are irreducible loci which are contained in the boundary, while other ones are the closures of irreducible loci in M*g*. But it is possible that some loci which are disjoint in M*g* have closures that meet in ¯M*g*.

We end this section with an explicit example, due to Antonio F. Costa, Milagros Izquierdo, and Hugo Parlier (cf. [129]), but we give a different proof. We refer to the notation of Example 204 in the following theorem.

Theorem 211 The strata *Na* and *N3+a* of M*g*(Z/3) fulfill:

(i) $N_a \cap N_{3+a} = \emptyset$

(ii $\overline{N_a} \cap \overline{N_{3+a}} \neq \emptyset \ in \ \overline{\mathfrak{M}_g}.$

Proof By the fact that inside Kuranishi space the locus of curves with a given action of a group *G* is a smooth manifold, hence locally irreducible, one sees right away that these strata must intersect in a point corresponding to a curve with at least two different automorphisms of order 3. A natural choice is to take a stable curve with an action of (Z/3)2.

The simplest choice is to take first the Fermat cubic

$$F := \left\{(x_0, x_1, x_2)|x_0^3 + x_1^3 + x_2^3 = 0\right\}$$

with the action of $(u, v) \in (\mathbb{Z}/3)^2$

$$(x_0, x_1, x_2) \mapsto (x_0, \epsilon^u x_1, \epsilon^v x_2).$$

We let then *C* be a (Z/3)-covering of the projective line, of genus *g*−3 and with ν(1)=*a*. Denote by *ψ* the covering automorphism and take a point *p*∈*C* which is not a ramification point: then glue the three points *ψv*(*p*), *v*∈Z/3, with the three points (1,0,−*ϵv*). In this way we get a stable curve *D* of genus *g*, and with an action of (Z/3)2.

We have two actions of Z/3 on *D*, the first induced by the action of (0, 1) on *F* and of *ψ* on *C*, and the second induced by the action of (1, 1) on *F* and *ψ* on *C* (it is immediate to observe that the glueing is respected by the action).

Now, for the first action of the generator ¯1∈Z/3 on *F*, the fixed points are the points *x*2=0, with local coordinate *t*=*x*2*x*0, such that *t*↦*ϵt*, while for the second action the fixed points are the points *x*0=0, with local coordinate *t*=*x*0*x*1, such that *t*↦*ϵ*2*t*.

Hence the same curve has two automorphisms of order 3, the first with ν(1)=*a*+3, the second with ν(1)=*a*. As easily shown (cf. [113]), since, for any of these two actions of Z/3, the nodes are not fixed, then there exists a smoothing for both actions of Z/3 on the curve D. This shows that the strata Na and $Na+3$ have D in their closure.

11.9 Branch stabilization and relation with other approaches

When $g'=0$ our $G\Gamma$ is related to the group $\hat{G}$ defined in [170] (Appendix), where the authors give a proof of a theorem by Conway and Parker. Roughly speaking the theorem says that: if the Schur multiplier $M(G)$ (which is isomorphic to $H2(G,Z)$) is generated by commutators, then the ν-type is a fine stable invariant, when $g'=0$.

Theorem 212 (Conway–Parker, loc.cit.) In the case $g'=0$, let $G=F/R$ where F is free, and assume that $H_2(G,\mathbb{Z}) \cong \frac{[F,F]\cap R}{[F,R]}$ is generated by commutators. Then there is an integer N such that if the numerical function ν takes values $\geq N$, then there is only one equivalence class with the given numerical function ν.

Michael Lönne, Fabio Perroni and myself have been able to extend this result in the following way:

Theorem 213 Assume that $g'=0$, or more generally that the union Γ of nontrivial conjugacy classes of G generates G.

Then there exists an integer N, depending on Γ, such that if two Hurwitz generating systems v, w satisfy $\nu(v)\geq N\chi\Gamma,\nu(w)\geq N\chi\Gamma$, where $\chi\Gamma$ is the characteristic function of the set of conjugacy classes corresponding toΓ, the Nielsen functions $\nu(v)$ and $\nu(w)$ are equivalent $\Leftrightarrow$ v is equivalent to w and they yield the same point in $A(g',d,G)$.

It is still an open question how to extend the previous result without the condition that Γ generates G.

Similar results have been obtained by Kulikov and Kharlamov [240], who use geometric arguments for the construction of semigroups similar to the ones constructed by Fried and Völklein, while a purely algebraic construction of a group similar to our group $K\Gamma$ can be found in work of Moravec on unramified Brauer groups ([297]).

11.10 Miller's description of the second homology of a group and developments

Clair Miller found ([286]) another nice description of the second homology group $H2(G,Z)$, as follows.

Definition 214 Let $\langle G,G\rangle$ be the free group on all pairs $\langle x,y\rangle$ with $x,y\in G$.

Then there is a natural surjection of $\langle G,G\rangle$ onto the commutator subgroup $[G, G]$ sending $\langle x,y\rangle$ to the commutator $[x,y]$.

Denote as in [286] by $Z(G)$ the kernel of this surjection (it might have been better to denote it by $Z(\langle G,G\rangle)$), and then denote by $B(G)$ (it might have been better to denote it by $B(\langle G,G\rangle)$) the normal subgroup of $Z(G)$ normally generated by the following elements

(1)$\langle x,y\rangle$

(2)$\langle x,y\rangle\langle y,x\rangle$

(3)$\langle z,xy\rangle\langle yx,zx\rangle\langle x,z\rangle$, where $yx:=xyx-1$,

(4)$\langle zx,yx\rangle\langle x,[y,z]\rangle\langle y,z\rangle$.

Theorem 215 (Miller) There is a canonical isomorphism $Z(G)/B(G) \cong H_2(G,\mathbb{Z})$.

Clearly we have then an exact sequence

$$0 \to H_2(G,\mathbb{Z}) \cong Z(G)/B(G) \to \langle G,G\rangle \to [G,G] \to 0$$

which corresponds to the previously seen

$$0 \to H_2(G,\mathbb{Z}) = (R\cap[F,F])/[F,R] \to [F,F]/[F,R] \to [G,G] \to 0.$$

The relations by Miller were later slightly modified by Moravec [297] in a more symmetric fashion as follows:

(1)$\langle g,g\rangle\sim 1$

(2)$\langle g1g2,h\rangle\sim\langle gg12,hg1\rangle\langle g1,h\rangle$

(3)$\langle g,h1h2\rangle\sim\langle g,h1\rangle\langle gh1,hh12\rangle$.

In the case where we consider also a finite union Γ of conjugacy classes, Moravec defined the following group

Definition 216 (*Moravec*)

$$G \wedge_{\Gamma} G = \langle G, G \rangle / B_{\Gamma},$$

where *B*Γ is defined by the previous relation (1), (2), (3) and by the further relation

$$(4)\ \langle g, k \rangle \sim 1, \quad \forall k \in \Gamma.$$

The definition by Moravec and the one given in [116] are related, as the following easy proposition shows.

Proposition 217 There is an exact sequence

$$0 \to H_{2,\Gamma}(G, \mathbb{Z}) \to G \wedge_{\Gamma} G \to [G, G] \to 0.$$

Another important source of interest for these quotient groups of *H*2(*G*,Z) comes from rationality questions.

In general, the stable cohomology of a finite group *G*, or more generally of an algebraic group *G*, is obtained in the following way.

Let *V* be a finite dimensional representation of *G*, and let *U* be an open set of *V* where *G* acts freely: then*U* / *G* is considered as an algebraic approximation to a classifying space for *G*, and one can take the limit over such representations (*V* and *V*′ begin smaller than their direct sum) of the cohomology groups of *U* / *G*. The same can of course be done also for homology groups and Chow groups (see [355]).

The groups that are thus obtained are important to study the problem of rationality of the quotients *V* / *G*, or of their stable rationality (*X* is said to be stably rational if there is an integer *n* such that *X*×P*n* is rational).

We refer to [20, 50, 51, 297] and the literature cited there for more details.

12 CONNECTED COMPONENTS OF MODULI SPACES AND THE ACTION OF THE ABSOLUTE GALOIS GROUP

Let X be a complex projective variety: let us quickly recall the notion of a conjugate variety.

Remark *218(1)ϕ∈Aut*(C) acts on C[*z*0,...*zn*], by sending $P(z) = \sum_{i=0}^{n} a_i z^i \mapsto \phi(P)(z) := \sum_{i=0}^{n} \phi(a_i) z^i$.

*(2)*Let X be as above a projective variety

$$X \subset \mathbb{P}^n_{\mathbb{C}},\ X := \{z | f_i(z) = 0 \quad \forall i\}.$$

The action of ϕ extends coordinatewise to $\mathbb{P}^n_{\mathbb{C}}$, and carries X to another variety, denoted $X\phi$, and called the**conjugate variety**. Since $fi(z)$=0 implies $\phi(fi)(\phi(z))$=0, we see that

$$X^\phi = \{w | \phi(f_i)(w) = 0 \quad \forall i\}.$$

If ϕ is complex conjugation, then it is clear that the variety $X\phi$ that we obtain is diffeomorphic to X; but, in general, what happens when ϕ is not continuous?

Observe that, by the theorem of Steiniz, one has a surjection *Aut*(C)→*Gal*(¯Q/Q), and by specialization the heart of the question concerns the action of *Gal*(¯Q/Q) on varieties X defined over ¯Q.

For curves, since in general the dimensions of spaces of differential forms of a fixed degree and without poles are the same for $X\phi$ and X, we shall obtain a curve of the same genus, hence $X\phi$ and X are diffeomorphic.

12.1 Galois conjugates of projective classifying spaces

General questions of which the first is answered in the positive in most concrete cases, and the second is answered in the negative in many cases, as we shall see, are the following.

Question 219 Assume that X is a projective $K(\pi,1)$, and assume $\phi \in Aut(\mathbb{C})$.

(A)Is then the conjugate variety $X\phi$ still a classifying space $K(\pi',1)$?

(B)Is then $\pi_1(X^\phi) \cong \pi \cong \pi_1(X)$?

Since ϕ is never continuous, there would be no reason to expect a positive answer to both questions (A) and (B), except that Grothendieck showed [201], see also [202].

Theorem 220 Conjugate varieties $X, X\phi$ have isomorphic algebraic fundamental groups

$$\pi_1(X)^{alg} \cong \pi_1(X^\phi)^{alg},$$

where $\pi 1(X)alg$ is the profinite completion of the topological fundamental group $\pi 1(X)$.

We recall once more that the profinite completion of a group G is the inverse limit

$$\hat{G} = lim_{K \trianglelefteq_f G}(G/K),$$

of the factor groups G / K, K being a normal subgroup of finite index in G; and since finite index subgroups of the fundamental group correspond to finite unramified (étale) covers, Grothendieck [201] defined in this way the algebraic fundamental group for varieties over other fields than the complex numbers, and also for more general schemes.

The main point of the proof of the above theorem is that if we have $f: Y \to X$ which is étale, also the Galois conjugate $f\phi: Y\phi \to X\phi$ is étale ($f\phi$ is just defined taking the Galois conjugate of the graph of f, a subvariety of $Y \times X$).

Since Galois conjugation gives an isomorphism of natural cohomology groups, which respects the cup product, as for instance the Dolbeault cohomology groups $H^p(\Omega^q_X)$, we obtain interesting consequences in the direction of question A) above. Recall the following definition.

Definition 221 Two varieties X, Y are said to be isogenous if there exist a third variety Z, and étale finite morphisms $f_X:Z\to X, f_Y:Z\to Y$.

Remark 222 It is obvious that if X is isogenous to Y, then X^ϕ is isogenous to Y^ϕ.

Theorem 223 *(i)*If X is an Abelian variety, or isogenous to an Abelian variety, the same holds for any Galois conjugateX^ϕ.

(ii)If S is a Kodaira fibred surface, then any Galois conjugate S^ϕ is also Kodaira fibred.

(iii)If X is isogenous to a product of curves, the same holds for any Galois conjugate X^ϕ.

Proof (i) X is an Abelian variety if and only it is a projective variety and there is a morphism $X\times X\to X$, $(x,y)\mapsto(x\cdot y^{-1})$, which makes X a group (see [303], it follows indeed that the group is commutative). This property holds for X if and only if it holds for X^ϕ.

(ii) The hypothesis is that there is $f:S\to B$ such that all the fibres are smooth and not all isomorphic: obviously the same property holds, after Galois conjugation, for $f^\phi:S^\phi\to B^\phi$.

(iii) It suffices to show that the Galois conjugate of a product of curves is a product of curves. But since $X^\phi\times Y^\phi=(X\times Y)^\phi$ and the Galois conjugate of a curve C of genus g is again a curve of the same genus g, the statement follows. ◻

Proceeding with other projective $K(\pi,1)$'s, the question becomes more subtle and we have to appeal to a famous theorem by Kazhdan on arithmetic varieties (see [117, 118, 235, 236, 287, 362]).

Theorem 224 Assume that X is a projective manifold with K_X ample, and that the universal covering $\tilde{X}$ is a bounded symmetric domain.

Let $\tau\in\mathrm{Aut}(\mathbb{C})$ be an automorphism of $\mathbb{C}$.

Then the conjugate variety X^τ has universal covering $\widetilde{X^\tau}\cong\tilde{X}$.

Simpler proofs follow from recent results obtained together with Antonio Di Scala, and based on the Aubin–Yau theorem and the results of Berger [39]. These results yield a precise characterization of varieties possessing a bounded symmetric domain as universal cover, and can be rather useful in view of the fact that our knowledge and classification of these fundamental groups is not so explicit.

To state them in detail would require some space, hence we just mention the simplest result (see [117]).

Theorem 225 Let X be a compact complex manifold of dimension n with KX ample.

Then the following two conditions (1) and (1'), resp. (2) and (2') are equivalent:

(1)*X* admits a slope zero tensor $0\neq\psi\in H0(Smn(\Omega 1X)(-mKX))$, (for some positive integer m);

(1')*X*$\cong\Omega/\Gamma$, where Ω is a bounded symmetric domain of tube type and Γ is a cocompact discrete subgroup of Aut(Ω) acting freely.

(2)*X* admits a semi special tensor $0 \neq \phi \in H^0(S^n(\Omega^1_X)(-K_X) \otimes \eta)$,, where η is a 2-torsion invertible sheaf, such that there is a point $p\in X$ for which the corresponding hypersurface $F_p := \{\phi_p = 0\} \subset \mathbb{P}(TX_p)$ is reduced.

(2')The universal cover of *X* is a polydisk.

Moreover, in case (1), the degrees and the multiplicities of the irreducible factors of the polynomial ψpdetermine uniquely the universal covering ˜$X=\Omega$.

12.2 Connected components of Gieseker's moduli space

For the sake of simplicity we shall describe in this and the next subsection the action of the absolute Galois group on the set of connected components of the moduli space of surfaces of general type. We first recall the situation concerning these components.

As we saw, all 5-canonical models of surfaces of general type with invariants $K2$, χ occur in a big family parametrized by an open set of the Hilbert scheme H0 parametrizing subschemes with Hilbert polynomial $P(m) = \chi + \frac{1}{2}(5m-1)5mK^2$, namely the open set

$$\mathcal{H}^0(\chi, K^2) := \{\Sigma | \Sigma \text{ is reduced with only R.D.P.'s as singularities } \}.$$

We shall however, for the sake of brevity, talk about connected components $\mathcal{N}$ of the Gieseker moduli space $\mathfrak{M}_{a,b}$ even if these do not really parametrize families of canonical models.

We refer to [108] for a more ample discussion of the basic ideas which we are going to sketch here.

$\mathfrak{M}_{a,b}$ has a finite number of connected components, and these parametrize the deformation classes of surfaces of general type with numerical invariants $\chi(S) = a,\ K_S^2 = b$. By the classical theorem of Ehresmann [148], deformation equivalent varieties are diffeomorphic, and moreover, via a diffeomorphism carrying the canonical class to the canonical class.

Hence, fixed the two numerical invariants $\chi(S)=a, K^2_S=b$, which are determined by the topology of S (indeed, by the Betti numbers $b_i(S)$ of S and by b^+:= positivity of the intersection form on $H^2(S,\mathbb{R})$), we have a finite number of differentiable types.

For some time the following question was open: whether two surfaces which are orientedly diffeomorphic would belong to the same connected component of the moduli space.

I conjectured (in [232]) that the answer should be negative, on the basis of some families of simply connected surfaces of general type constructed in [82] and later investigated in [84, 104] and [111]: these were shown to be homeomorphic by the results of Freedman (see [166, 167]), and it was then relatively easy to show then [85] that there were many connected components of the moduli space corresponding to homeomorphic but non diffeomorphic surfaces. It looked like the situation should be similar even if one would fix the diffeomorphism type.

Friedman and Morgan instead made the 'speculation' that the answer to the DEF = DIFF question should be positive (1987) (see [168]), motivated by the new examples of homeomorphic but not diffeomorphic surfaces discovered by Donaldson (see [138] and [139] for a survey on this topic).

The question was finally answered in the negative, and in every possible way ([26, 98, 104, 239, 278]).

Theorem 226 (Manetti '98, Kharlamov–Kulikov 2001, C. 2001, C.-Wajnryb 2004, Bauer–C.-Grunewald 2005)

The Friedman–Morgan speculation does not hold true and the DEF = DIFF question has a negative answer.

In my joint work with Wajnryb [104] the DEF = DIFF question was shown to have a negative answer also for simply connected surfaces (indeed for some of the families of surfaces constructed in [82]).

I refer to [106] for a rather comprehensive treatment of the above questions (and to [3, 17, 100, 107, 111] for the symplectic point of view, [37, 137] for the special case of geometric genus *pg*= 0).

12.3 Arithmetic of moduli spaces and faithful actions of the absolute Galois group

A basic remark is that all the schemes involved in the construction of the Gieseker moduli space are defined by equations involving only Z-coefficients, since the defining equation of the Hilbert scheme is a rank condition for a multiplication map (see for instance [188]), and similarly the condition $\omega_\Sigma^{\otimes 5} \cong \mathcal{O}_\Sigma(1)$ is also closed (see [303]) and defined over Z.

It follows that the absolute Galois group *Gal*(¯Q,Q) acts on the Gieseker moduli space M*a,b*. In particular, it acts on the set of its irreducible components, and on the set of its connected components.

After an incomplete initial attempt in [28] in joint work with Ingrid Bauer and Fritz Grunewald, we were able in [29] to show:

Theorem 227 The absolute Galois group *Gal*(¯Q/Q) acts faithfully on the set of connected components of the Gieseker moduli space of surfaces of general type,

$$\mathfrak{M} := \cup_{x,y \in \mathbb{N}, x,y \geq 1} \mathfrak{M}_{x,y}.$$

Another result in a similar direction had been obtained by Easton and Vakil [143] using abelian coverings of the plane branched on union of lines.

Theorem 228 The absolute Galois group $Gal(\bar{\mathbb{Q}}/\mathbb{Q})$ acts faithfully on the set of irreducible components of the Gieseker moduli space of surfaces of general type,

$$\mathfrak{M} := \cup_{x,y\in\mathbb{N},x,y\geq 1}\mathfrak{M}_{x,y}.$$

The main ingredients for the proof of Theorem 227 are the following ones.

(1)Define, for any complex number $a\in \mathbb{C}\setminus\{-2g,0,1,\ldots,2g-1\}$, C_a as the hyperelliptic curve of genus $g\geq 3$ which is the smooth complete model of the affine curve of equation

$$w^2 = (z-a)(z+2g)\Pi_{i=0}^{2g-1}(z-i).$$

Consider then two complex numbers a, b such that $a\in \mathbb{C}\setminus\mathbb{Q}$: then $C_a\cong C_b$ if and only if $a=b$.

(2)If $a\in\bar{\mathbb{Q}}$, then by Belyi's theorem [38] there is a morphism $f_a:C_a\to\mathbb{P}^1$ which is branched only on three points, $0,1,\infty$.

(3)The normal closure D_a of f_a yields a triangle curve, i.e., a curve D_a with the action of a finite group $G_a$such that $D_a/G_a\cong\mathbb{P}^1$, and $D_a\to\mathbb{P}^1$ is branched only on three points.

(4)Take surfaces isogenous to a product $S=(D_a\times D')/G_a$ where the action of G_a on D' is free. Denote by $\mathcal{N}_a$ the union of connected components parametrizing such surfaces.

(5)Take all the possible twists of the G_a-action on $D_a\times D'$ via an automorphism $\psi\in Aut(G_a)$ (i.e., given the action $(x,y)\mapsto(\gamma x,\gamma y)$, consider all the actions of the form

$$(x,y)\mapsto(\gamma x,\psi(\gamma)y).$$

One observes that, for each ψ as above, we get more connected components in $\mathcal{N}_a$.

(6)Find by an explicit calculation (using (4) and (5)) that the subgroup of $Gal(\bar{\mathbb{Q}}/\mathbb{Q})$ acting trivially on the set of connected components of the moduli space would be a normal and abelian subgroup.

(7)Finally, this contradicts a known theorem (cf. [169]).

Remark 229 An interesting remark of a referee is that the meaning of the above theorem could be further elaborated as a parallel of the Drinfend theory of Galois representation, built on the theorem of Belyi (see [140]). We leave this as a future task.

12.4 Change of fundamental group

Jean Pierre Serre proved in the 60's [336] the existence of a field automorphism $\phi \in Gal(\bar{Q}/Q)$, and a variety$X$ defined over $\bar{Q}$ such that X and the Galois conjugate variety $X\phi$ have non isomorphic fundamental groups.

In [29] this phenomenon is vastly generalized, thus answering question B) in the negative.

Theorem 230 If $\sigma \in Gal(\bar{Q}/Q)$ is not in the conjugacy class of complex conjugation, then there exists a surface isogenous to a product X such that X and the Galois conjugate surface $X\sigma$ have non-isomorphic fundamental groups.

Since the argument for the above theorem is not constructive, let us observe that, in work in collaboration with Bauer and Grunewald [27, 29], we discovered wide classes of explicit algebraic surfaces isogenous to a product for which the same phenomenon holds.

By the strong rigidity of locally symmetric spaces X=D/Γ whose universal covering D is an irreducible bounded symmetric domain of dimension ≥2, similar phenomena should also occur in this case.

Remark 231 Further developments have been announced in [183] by Gonzaléz-Diez and Jaikin-Zapirain: for instance the faithfulness of the action of the absolute Galois group on the discrete set of the moduli space corresponding to Beauville surfaces, and the extension of Theorem 230 to all automorphisms σ different from complex conjugation.

13 STABILIZATION RESULTS FOR THE HOMOLOGY OF MODULI SPACES OF CURVES AND ABELIAN VARIETIES

We have seen that the moduli space of curves is a rational $K(\pi,1)$, since it can be written as a quotient of the Teichmüller space T*g* of a closed oriented real 2-manifold *M* of genus *g*

$$\mathfrak{M}_g = \mathcal{T}_g/Map_g, \quad \mathcal{T}_g := \mathcal{CS}(M)/\mathrm{Diff}^0(M),$$

As a corollary, and as we saw, the rational cohomology of the moduli space is calculated by group cohomology:

$$H^*(\mathfrak{M}_g, \mathbb{Q}) \cong H^*(\mathrm{Map}_g, \mathbb{Q}).$$

Harer showed, using the concept of genus stabilization that we have already introduced in Sect. 10, that these cohomology groups stabilize with *g*; indeed, stabilization furnishes an inclusion of a space which is homotopically equivalent to M*g* inside M*g*+1 (alternatively, one may say that *Mapg*→*Mapg*+1, by letting the operation be trivial on the added handle).

Theorem 232 (Harer [208]) Let $\mathrm{Map}^s_{g,r}$ be the mapping class group of an orientable surface *F* of genus *g* with *r* boundary components and *s* punctures. Then, for $g \geq 3k - 1$, $H_k(\mathrm{Map}^s_{g,r}, \mathbb{Z})$ is independent of *g* and *r* as long as *r*>0, for $g \geq 3k$, $H_k(\mathrm{Map}^s_{g,r}, \mathbb{Q})$ is independent of *g* and *r* for every *r*, and for *g*≥3*k*+1, *Hk*(Maps*g*,*r*,Z) is independent of *g* and *r* for every *r*.

The ring structure of the cohomology of the "stable mapping class group" is described by a conjecture of Mumford [307], that has been proven by Madsen and Weiss [274].

Theorem 233(Mumford's conjecture) The stable cohomology of the moduli space of curves is a polynomial algebra

$$H^*(Map_\infty, \mathbb{Q}) = \mathbb{Q}[\mathcal{K}_1, \mathcal{K}_2, \ldots]$$

where the class K*i* is the direct image (*pg*)∗(*Ki*+1) of the (*i*+1)-th power of the relative canonical divisor of the universal family of curves.

These results paralleled earlier results of Borel [58] and Charney and Lee [120] on the cohomology of arithmetic varieties, such as the moduli space of Abelian varieties and some partial compactifications of them.

For instance, in the case of the moduli space of Abelian varieties we have the following theorem by Borel.

Theorem 234 (Borel) The stable cohomology of the moduli space of Abelian varieties is a polynomial algebra

$$H^*(\mathcal{A}_\infty, \mathbb{Q}) = \mathbb{Q}[\lambda_1, \lambda_3, \lambda_5 \ldots]$$

where the class λi is the i-th Chern class of the universal bundle $H$1,0.

The theme of homology (cohomology-) stabilization is indeed a very general one, which has been recently revived through the work of several authors, also in other contexts (see for instance [131, 153, 158, 203]).

It would take too long to dwell here on this topic, which would deserve a whole survey article devoted to it.

13.1 Epilogue

There are many other interesting topics which are very tightly related to the main theme of this article.

For instance, there is a relation between symmetric differentials and the fundamental group ([52, 53, 64,241]).

Brunebarbe, Klingler and Totaro showed indeed in [64] that some 'linearity' property of the fundamental group entails the existence of nontrivial symmetric differentials.

Theorem 235 Let X be a compact Kähler manifold, and let k be any field.

Assume that there is a representation

$$\rho : \pi_1(X) \to GL(r, k)$$

with infinite image.

Then the symmetric algebra of X is nontrivial,

$$\oplus_{m\geq 0} H^0(S^m(\Omega^1_X)) \neq \mathbb{C}.$$

Another very interesting topic is Gromov's h-principle, for which we refer the reader to [152], and [74]. Perhaps not only the author is tired at this point.

FOOTNOTES

1In these days the angel of topology and the devil of abstract algebra fight for the soul of each individual mathematical domain [373], p. 500.

My motto is instead: 'Any good mathematical theory requires several good theorems. Conversely, a really good theorem requires several good theories. '

2One uses here the following trick: the intersection of a projective variety X with a hypersurface is equal to the intersection of X with a hyperplane, but for a different embedding of X; this trick is used several times, starting with X=PN.

3The reduced homology group differs from the ordinary one only for i=0, and for i=0 is defined as the kernel of the degree surjection onto Z: this distinction is only needed in order to treat the case n=1 on an equal footing.

4Following a suggestion of Marco Manetti.

5The same result holds without the assumption that ρ be injective, since it suffices to consider the image $G'=Im(\rho)$, apply the pull-back construction to get $1\to\pi g\to\hat{}G'\to G'\to 1$, and then construct $\hat{}G$ as the inverse image of the diagonal under the epimorphism $\hat{}G'\times G\to G'\times G'$.

6One can indeed relax the hypothesis, requiring only that X is normal and Y is smooth.

7As we shall see in the last sections of the article, the absolute Galois group permutes the connected components of the moduli space of surfaces of general type, and this action is indeed faithful.

8The flow method made then its way further through the work of Hamilton [206], Perelman and others [320–322], leading to the solution of the three dimensional Poincaré conjecture (see for example [299] for an exposition).

9Hence also of $SL(2,C)$.

10This last property for algebraic surfaces follows automatically from homotopy invariance.

11In [35, 37] we made this assertion for any finite group G, see however Remark 23.

12This means: there exists an algebraic family of surfaces $S=\cup_{t\in T}S_t$ over a smooth curve T, and an algebraic family of maps $\varphi_t:S_t\to S_t, t\in T$, i.e., such that the union of the graphs of the φ_t's is a closed algebraic set in the fibre product $S\times_T S$, with the property that φ_t is an isomorphism exactly for $t\neq t_0$.

ACKNOWLEDGMENTS

Thanks to Efim Zelmanov for the honour of inviting me to write this article, and for his encouragement and support during its preparation. Thanks also to Maurizio Cornalba: some parts of this survey, which are directed to a rather general public, are influenced by an unpublished draft which I wrote after a plenary lecture which I gave at the XIV Congress of the Italian Mathematical Union (Catania 1991). Thanks to Ingrid Bauer, Michael Lönne and Fabio Perroni, for the pleasure I got from collaborating with them and for their invaluable contributions to our exciting joint research. A substantial part of the paper was written when I was visiting KIAS as research scholar: I am very grateful for the excellent atmosphere and mathematical environment I found at KIAS. Lectures held in Bayreuth, KAIST Daejeon and Centre Henri Lebesgue in Angers were also very useful. Support of the Forschergruppe 790 'Classification of algebraic surfaces and compact complex manifolds' of the DFG, and of the ERC Advanced Grant No. 340258, 'TADMICAMT' is gratefully acknowledged. Finally, thanks to the students and postdocs who listened to my lectures in Bayreuth for their remarks, and to Ingrid Bauer, Binru Li, Sascha Weigl, and especially Wenfei Liu, for pointing out misprints and corrections to be made.

REFERENCES

1. Alexandrov, P.: Introduction á la Théorie homologique de la dimension, French translation of the original Russian volume of 1975. Editions MIR, Moscow (1977)

2. Amorós, J., Burger, M., Corlette, A., Kotschick, D., Toledo, D.: Fundamental Groups of Compact Kähler Manifolds. Mathematical Surveys and Monographs. 44. Providence, RI: American Mathematical Society (AMS). xi, p. 140 (1996)

3. Amoros, J., Bogomolov, F., Katzarkov, L., Pantev, T.: Symplectic Lefschetz fibrations with arbitrary fundamental group, with an appendix by Ivan Smith. J. Differ. Geom. 54(3), 489–545 (2000)

4. Andreotti, A., Frankel, T.: The Lefschetz theorem on hyperplane sections. Ann. Math. 2(69), 713–717 (1959)

5. Andreotti, A., Frankel, T.: The second Lefschetz theorem on hyperplane sections. In: Spencer, D.C., Iyanaga, S. (eds.) Global Analysis (Papers in Honor of K. Kodaira). Univ. Tokyo Press, Tokyo, pp. 1–20 (1969)

6. Andreotti, A., Grauert, H.: Théoréme de finitude pour la cohomologie des espaces complexes. Bull. Soc. Math. France 90, 193–259 (1962)

7. Arapura, D., Bressler, P., Ramachandran, M.: On the fundamental group of a compact Kähler manifold. Duke Math. J. 68(3), 477–488 (1992)

8. Arbarello, E., Cornalba, M.: Teichmüller space via Kuranishi families. Ann. Sc. Norm. Super. Pisa Cl. Sci. (5) 8(1), 89–116 (2009)

9. Artin, E.: Theorie der Zöpfe. Hamburg Univ. Math. Seminar Abhandlungen 4–5, 47–72 (1926)

10. Artin, E.: The collected papers of Emil Artin, edited by Serge Lang and John T. Tate, Addison-Wesley Publishing Co., Inc., Reading, Mass.- London (1965)

11. Artin, M.: Versal deformations and algebraic stacks. Invent. Math. 27, 165–189 (1974)

12. Atiyah, M.F.: Vector bundles over an elliptic curve. Proc. Lond. Math. Soc. 3(7), 414–452 (1957)

13. Atiyah, M.F.: On analytic surfaces with double points. Proc. R. Soc. Lond. Ser. A 247, 237–244 (1958)

14. Atiyah, M.F., Hitchin, N.J., Singer, I.M.: Self-duality in four-dimensional Riemannian geometry. Proc. R. Soc. Lond. Ser. A 362(1711), 425–461 (1978)

15. Atiyah, M.F., Bott, R.: The Yang-Mills equations over Riemann surfaces. Philos. Trans. R. Soc. Lond. A308, 523–615 (1983)

16. Aubin, T.: Équations du type Monge-Ampére sur les variétés kählériennes compactes. Bull. Sci. Math. (2) 102(1), 63–95 (1978)

17. Auroux, D., Katzarkov, L.: Branched coverings of CP2 and invariants of symplectic 4-manifolds. Invent. Math. 142, 631–673 (2000)

18. Badescu, L.: Projective geometry and formal geometry. Instytut Matematyczny Polskiej Akademii Nauk. Monografie Matematyczne (New Series), 65. Birkhäuser Verlag, Basel, pp. xiv+209 (2004)

19. Bagnera, G., de Franchis, M.: Le superficie algebriche le quali ammettono una rappresentazione parametrica mediante funzioni iperellittiche di due argomenti. Mem. di Mat. e di Fis. Soc. It. Sc. (3) 15, 253–343 (1908)

20. Baranovsky, P., Petrov, T.: Brauer groups and crepant resolutions. Adv. Math. 209, 547–560 (2007)

21. Barth, W., Peters, C., Van de Ven, A.: Compact complex surfaces. Ergebnisse der Mathematik und ihrer Grenzgebiete (3), 4. Springer-Verlag, Berlin (1984); 2nd edition by W. Barth, K. Hulek, C. Peters, A. Van de Ven, Ergebnisse der Mathematik und ihrer Grenzgebiete. 3. Folge. A, 4. Springer-Verlag, Berlin (2004)

22. Barth, W.: Transplanting cohomology classes in complex-projective space. Am. J. Math. 92, 951–967 (1970)

23. Barth, W., Larsen, M.E.: On the homotopy groups of complex projective algebraic manifolds. Math. Scand.30, 88–94 (1972)

24. Barth, W., Van de Ven, A.: A decomposability criterion for algebraic 2-bundles on projective spaces. Invent. Math. 25, 91–106 (1974)

25. Bauer, I.: Irrational pencils on non-compact algebraic manifolds. Int. J. Math. 8(4), 441–450 (1997)

26. Bauer, I., Catanese, F., Grunewald, F.: Beauville surfaces without real structures. In: Geometric methods in algebra and number theory. Progr. Math. 235, 1–42, Birkhäuser (2005)

27. Bauer, I., Catanese, F., Grunewald, F.: Chebycheff and Belyi polynomials, dessins d'enfants, Beauville surfaces and group theory. Mediterr. J. Math. 3(2), 119–143 (2006)

28. Bauer, I., Catanese, F., Grunewald, F.: The absolute Galois group acts faithfully on the connected components of the moduli space of surfaces of general type, p. 13. arXiv:0706.1466

29. Bauer, I., Catanese, F., Grunewald, F.: Faithful actions of the absolute Galois group on connected components of moduli spaces. Invent. Math. 199, 859–888 (2015) (published online in 2014). http://link.springer.com/article/10.1007/s00222-014-0531-2

30. Bauer, I., Catanese, F.: The moduli space of Keum-Naie surfaces. Groups Geom. Dyn. 5(2), 231–250 (2011)

31. Bauer, I., Catanese, F.: Burniat surfaces I: fundamental groups and moduli of primary Burniat surfaces. In: Faber, C et al. (ed.) Classification of algebraic varieties. Based on the conference on classification of varieties. Schiermonnikoog, Netherlands (2009). Zürich: European Mathematical Society (EMS). EMS Series of Congress Reports, pp. 49–76 (2011)

32. Bauer, I., Catanese, F.: Burniat surfaces. II. Secondary Burniat surfaces form three connected components of the moduli space. Invent. Math. 180(3), 559–588 (2010)

33. Bauer, I., Catanese, F.: Burniat surfaces III: deformations of automorphisms and extended Burniat surfaces. Doc. Math. 18, 1089–1136 (2013)

34. Bauer, I., Catanese, F., Pignatelli, R.: Surfaces with geometric genus zero: a survey. Ebeling, W. et al. (ed.) Complex and differential geometry. Conference held at Leibniz Universität Hannover, Germany, September 14–18, 2009. Proceedings. Berlin, Springer. Springer Proceedings in Mathematics 8, 1–48 (2011)

35. Bauer, I., Catanese, F.: Inoue type manifolds and Inoue surfaces: a connected component of the moduli space of surfaces with K2=7,pg=0. Geometry and arithmetic. 23–56, EMS Ser. Congr. Rep., Eur. Math. Soc., Zürich (2012)

36. Bauer, I., Catanese, F.: Burniat-type surfaces and a new family of surfaces with pg=0,K2=3 Rend. Circ. Mat. Palermo (2) 62(1), 37–60 (2013)

37. Bauer, I., Catanese, F., Frapporti, D.: Generalized Burniat type surfaces and Bagnera-de Franchis varieties. J. Math. Sci. Univ. Tokyo 22, 55–111 (2015)

38. Belyĭ, G.V.: On Galois extensions of a maximal cyclotomic field. Izv. Akad. Nauk SSSR Ser. Mat. 43(2), 269–276 (1979). Translation in Math. USSR- Izv. 14, 247–256 (1980)

39. Berger, M.: Sur les groupes d'holonomie homogéne des variétés á connexion affine et des variétés riemanniennes. Bull. Soc. Math. France 83, 279–330 (1953)

40. Bers, L.: Simultaneous uniformization. Bull. Am. Math. Soc. 66, 94–97 (1960)

41. Bestvina, M., Church, T., Souto, J.: Some groups of mapping classes not realized by diffeomorphisms. Comment. Math. Helv. 88(1), 205–220 (2013)

42. Bierstone, E., Pierre, M.: Semianalytic and subanalytic sets. Publ. Math. Inst. Hautes Étud. Sci. 67, 5–42 (1988)

43. Birkar, C., Cascini, P., Hacon, C.D., McKernan, J.: Existence of minimal models for varieties of log general type. J. Am. Math. Soc. 23(2), 405–468 (2010)

44. Birman, J.S.: Mapping class groups and their relationship to braid groups. Commun. Pure Appl. Math. 22, 213–238 (1969)

45. Birman, J.S.: Braids, Links, and Mapping Class Groups. Annals of Mathematics Studies, No. 82. Princeton University Press, Princeton, NJ; University of Tokyo Press, Tokyo (1974)

46. Blanchard, A.: Recherche de structures analytiques complexes sur certaines variétés. C. R. Acad. Sci. Paris 236, 657–659 (1953)

47. Blanchard, A.: Espaces fibrés kählériens compacts. C. R. Acad. Sci. Paris 238, 2281–2283 (1954)

48. Blanchard, A.: Sur les variétés analytiques complexes. Ann. Sci. Ecole Norm. Sup. 3(73), 157–202 (1956)

49. Bogomolov, F.A., Katzarkov, L.: Complex projective surfaces and infinite groups. Geom. Funct. Anal. 8(2), 243–272 (1998)

50. Bogomolov, F.A.: Holomorphic tensors and vector bundles on projective manifolds. Izv. Akad. Nauk SSSR Ser. Mat. 42(6), 1227–1287, 1439 (1978)

51. Bogomolov, F.A.: The Brauer group of quotient spaces of linear representations. Math. USSR-Izv. 30(3), 455–485 (1988)

52. Bogomolov, F., De Oliveira, B.: Symmetric differentials of rank 1 and holomorphic maps. Pure Appl. Math. Q. 7(4), 1085–1103. Special Issue: In memory of Eckart Viehweg (2011)

53. Bogomolov, F., De Oliveira, B.: Closed symmetric 2-differentials of the 1st kind. Pure Appl. Math. Q. 9(4), 613–642 (2013)

54. Bombieri, E.: Canonical models of surfaces of general type. Publ. Math. I.H.E.S 42, 173–219 (1973)

55. Bombieri, E., Husemoller, D.: Classification and embeddings of surfaces. In: Algebraic geometry, Humboldt State Univ., Arcata, Calif., 1974. Proc. Sympos. Pure Math., vol. 29, Amer. Math. Soc., Providence, R.I., pp. 329–420 (1975)

56. Borel, A.: Seminar on transformation groups. With contributions by Bredon, G., Floyd, E.E., Montgomery, D., Palais, R. (eds.) Annals of Mathematics Studies, No. 46. Princeton University Press, Princeton, NJ, p. 245 (1960)

57. Borel, A.: Compact Clifford-Klein forms of symmetric spaces. Topology 2, 111–122 (1963)

58. Borel, A.: Stable real cohomology of arithmetic groups. Ann. Sci. École Norm. Sup. (4) 7 (1974) 235–272 (1975)

59. Borel, A., Moore, J.C.: Homology theory for locally compact spaces. Mich. Math. J. 7, 137–159 (1960)

60. Brieskorn, E.: Rationale Singularitäten komplexer Flächen, Invent. Math. 4, 336–358 (1967/1968)

61. Brieskorn, E.: Die Auflösung der rationalen Singularitäten holomorpher Abbildungen. Math. Ann. 178, 255–270 (1968)

62. Brieskorn, E.: Singular elements of semi-simple algebraic groups. 'Actes du Congrés International des Mathématiciens' (Nice, 1970) Tome 2. Gauthier-Villars, Paris, 279–284 (1971)

63. Brown, K.S.: Cohomology of Groups. Graduate Texts in Mathematics, 87. Springer-Verlag, New York (1982)

64. Brunebarbe, Y., Klingler, B., Totaro, B.: Symmetric differentials and the fundamental group. Duke Math. J.162(14), 2797–2813 (2013)

65. Burniat, P.: Sur les surfaces de genre P12>1. Ann. Mat. Pura Appl. 4(71), 1–24 (1966)

66. Burns, D., Wahl, J.: Local contributions to global deformations of surfaces. Invent. Math. 26, 67–88 (1974)

67. Cai, J.X.: Classification of fiber surfaces of genus 2 with automorphisms acting trivially in cohomology. Pac. J. Math. 232(1), 43–59 (2007)

68. Cai, J.-X., Liu, W., Zhang, L.: Automorphisms of surfaces of general type with q≥2 acting trivially in cohomology. Compos. Math. 149(10), 1667–1684 (2013)

69. Calabi, E.: Construction and properties of some 6-dimensional almost complex manifolds. Trans. Am. Math. Soc. 87, 407–438 (1958)

70. Calabi, E., Vesentini, E.: On compact, locally symmetric Kähler manifolds. Ann. Math. 2(71), 472–507 (1960)

71. Campana, F.: Remarques sur le revêtement universel des variétés kählériennes compactes. Bull. Soc. Math. France 122(2), 255–284 (1994)

72. Campana, F.: Kodaira dimension and fundamental group of compact Kähler manifolds. Lecture Notes Series, 7. Universitä di Trento, Dipartimento di Matematica (1995)

73. Campana, F.: Fundamental group and positivity of cotangent bundles of compact Kähler manifolds. J. Algebraic Geom. 4(3), 487–502 (1995)

74. Campana, F., Winkelmann, J.: On h-principle and specialness for complex projective manifolds. arXiv:1210.7369

75. Carlson, J.A., Toledo, D.: Harmonic mappings of Kähler manifolds to locally symmetric spaces. Inst. Hautes Études Sci. Publ. Math. 69, 173–201 (1989)

76. Cartan, E.: Lecons sur la géométrie des espaces de Riemann. Paris: Gauthier-Villars (Cahiers scientifiques publiés sous la direction de G. Julia, 2). VI, p. 273 (1928)

77. Cartan, E.: Sur les domaines bornés homogénes de l'espace des n variables complexes. Abhandl. Hamburg 11, 116–162 (1935)

78. Cartan, H.: Quotient d'un espace analytique par un groupe d'automorphismes. In: A symposium in honor of S. Lefschetz. Algebraic geometry and topology. Princeton University Press, Princeton, NJ, pp. 90–102 (1957)

79. Cartan, H., Eilenberg, S.: Homological Algebra. Princeton University Press, Princeton, NJ, pp. xv+390 (1956)

80. Castelnuovo, G.: Sulle superficie aventi il genere aritmetico negativo. Palermo Rend. 20, 55–60 (1905)

81. Catanese, F.: Moduli of surfaces of general type. In: Ciliberto, C., Ghione, E., Orecchia, E. (eds.) Algebraic geometry-open problems. (Ravello, 1982). Lecture Notes in Math., 997. Springer, Berlin-New York, pp. 90–112 (1983)

82. Catanese, F.: On the Moduli spaces of surfaces of general type. J. Differ. Geom. 19, 483–515 (1984)

83. Catanese, F.: Commutative algebra methods and equations of regular surfaces. In: Algebraic Geometry, Bucharest 1982 (Bucharest, 1982), pp. 68–111. Lecture Notes in Math., 1056, Springer, Berlin (1984)

84. Catanese, F.: Automorphisms of rational double points and moduli spaces of surfaces of general type. Compos. Math. 61, 81–102 (1987)

85. Catanese, F.: Connected components of moduli spaces. J. Differ. Geom. 24, 395–399 (1986)

86. Catanese, F.: Moduli of algebraic surfaces. Theory of moduli (Montecatini Terme, 1985). Lecture Notes in Math. 1337. Springer, Berlin, pp. 1–83 (1988)

87. Catanese, F.: Moduli and classification of irregular Kähler manifolds (and algebraic varieties) with Albanese general type fibrations. Invent. Math. 104, 263–289 (1991)

88. Catanese, F.: Chow varieties, Hilbert schemes and moduli spaces of surfaces of general type. J. Algebraic Geom. 1(4), 561–595 (1992)

89. Catanese, F., Ciliberto, C.: On the irregularity of cyclic coverings of algebraic surfaces. Geometry of complex projective varieties (Cetraro, 1990), pp. 89–115, Sem. Conf., 9, Mediterranean Press, Rende (1993)

90. Catanese, F.: Compact complex manifolds bimeromorphic to tori. In: Abelian varieties (Egloffstein, 1993). de Gruyter, Berlin, pp. 55–62 (1995)

91. Catanese, F.: Fundamental groups with few relations. In: Higher dimensional complex varieties. Proceedings of the international conference, Trento, Italy, June 15–24, 1994. Walter de Gruyter, Berlin, pp. 163–165 (1996)

92. Catanese, F.: (Some) old and new results on algebraic surfaces. First European Congress of Mathematics, vol. I (Paris, 1992), pp. 445–490, Progr. Math., 119, Birkhäuser, Basel (1994)

93. Catanese, F., Franciosi, M.: Divisors of small genus on algebraic surfaces and projective embeddings. In: Proceedings of the Hirzebruch 65 Conference on Algebraic Geometry (Ramat Gan, 1993). Israel Math. Conf. Proc., 9, Bar-Ilan Univ., Ramat Gan, pp. 109–140 (1996)

94. Catanese, F., Franciosi, M., Hulek, K., Reid, M.: Embeddings of curves and surfaces. Nagoya Math. J.154, 185–220 (1999)

95. Catanese, F.: Singular bidouble covers and the construction of interesting algebraic surfaces. In: Pragacz, P., Szurek, M.,

Wisniewski, J. (eds.) Algebraic Geometry: Hirzebruch 70, vol. 241. AMS Cont. Math., Providence, RI, pp. 97–120 (1999)

96. Catanese, F.: Fibred surfaces, varieties isogenous to a product and related moduli spaces. Am. J. Math.122(1), 1–44 (2000)

97. Catanese, F., Frediani, P.: Real hyperelliptic surfaces and the orbifold fundamental group. J. Inst. Math. Jussieu 2(2), 169–233 (2003)

98. Catanese, F.: Moduli spaces of surfaces and real structures. Ann. Math. (2) 158(2), 577–592 (2003)

99. Catanese, F.: Deformation types of real and complex manifolds. Contemporary trends in algebraic geometry and algebraic topology (Tianjin, 2000), 195238, Nankai Tracts Math., 5. World Sci. Publ., River Edge, NJ (2002)

100. Catanese, F.: Symplectic structures of algebraic surfaces and deformation, p. 14. math.AG/0207254

101. Catanese, F.: Fibred Kähler and quasi projective groups. Adv. Geom. suppl., Special Issue dedicated to A. Barlotti's 80-th birthday (2003), Adv. Geom. suppl. S13–S27 (2003)

102. Catanese, F., Keum, J., Oguiso, K.: Some remarks on the universal cover of an open K3 surface. Math. Ann. 325(2), 279–286 (2003)

103. Catanese, F.: Deformation in the large of some complex manifolds, I. Ann. Mat. Pura Appl. 183(4), Volume in Memory of Fabio Bardelli (3), 261–289 (2004)

104. Catanese, F., Wajnryb, B.: Diffeomorphism of simply connected algebraic surfaces. J. Differ. Geom.76(2), 177–213 (2007)

105. Catanese, F.: Surface classification and local and global fundamental groups. I. Atti Accad. Naz. Lincei Cl. Sci. Fis. Mat. Natur. Rend. Lincei (9) Mat. Appl. 17(2), 135–153 (2006)

106. Catanese, F.: Differentiable and deformation type of algebraic surfaces, real and symplectic structures. Symplectic 4-manifolds and

algebraic surfaces, 55–167. Lecture Notes in Math., 1938. Springer, Berlin (2008)

107. Catanese, F.: Canonical symplectic structures and deformations of algebraic surfaces. Commun. Contemp. Math. 11(3), 481–493 (2009)

108. Catanese, F.: Algebraic surfaces and their moduli spaces: real, differentiable and symplectic structures. Boll. Unione Mat. Ital. (9) 2(3), 537–558 (2009)

109. Catanese, F., Rollenske, S.: Double Kodaira fibrations. J. Reine Angew. Math. 628, 205–233 (2009)

110. Catanese, F., Oguiso, K., Peternell, T.: On volume-preserving complex structures on real tori. Kyoto J. Math. 50(4), 753–775 (2010)

111. Catanese, F., Lönne, M., Wajnryb, B.: Moduli spaces and braid monodromy types of bidouble covers of the quadric. Geom. Topol. 15, 351–396 (2011). doi:10.2140/gt.2011.15.351. URL: http://www.msp.warwick.ac.uk/gt/2011/15-01/p010.xhtml

112. Catanese, F., Lönne, M., Perroni, F.: Irreducibility of the space of dihedral covers of the projective line of a given numerical type. Atti Accad. Naz. Lincei Cl. Sci. Fis. Mat. Natur. Rend. Lincei (9) Mat. Appl. 22(3), 291–309 (2011)

113. Catanese, F.: Irreducibility of the space of cyclic covers of algebraic curves of fixed numerical type and the irreducible components of Sing(¯Mg). In: Janeczko, S., Li J., Phong, D.H. (eds) 'Advances in Geometric Analysis', in honor of Shing-Tung Yau's 60th birthday. Advanced Lectures in Mathematics 21, International Press (Somerville, USA) and Higher Education Press (Beijing, China), pp. 281–306 (2012)

114. Catanese, F.: A superficial working guide to deformations and moduli. In: Farkas, G., Morrison, I. (eds.) Handbook of moduli, vol. I,

pp. 161–215. Adv. Lect. Math. (ALM), 24, Int. Press, Somerville, MA (2013)

115. Catanese, F., Lönne, M., Perroni, F.: The irreducible components of the moduli space of dihedral covers of algebraic curves. arXiv:1206.5498 (to appear in Groups, Geometry and Dynamics)

116. Catanese, F., Lönne, M., Perroni, F.: Genus stabilization for moduli of curves with symmetries. arXiv:1301.4409 (to appear in Algebraic Geometry)

117. Catanese, F., Di Scala, A.J.: A characterization of varieties whose universal cover is the polydisk or a tube domain. Math. Ann. 356(2), 419–438 (2013)

118. Catanese, F., Di Scala, A.J.: A characterization of varieties whose universal cover is a bounded symmetric domain without ball factors. Adv. Math. 257, 567–580 (2014)

119. Chan, T.O.M., Coughlan, S.: Kulikov surfaces form a connected component of the moduli space. Nagoya Math. J. 210, 1–27 (2013)

120. Charney, R., Lee, R.: Cohomology of the Satake compactification. Topology 22(4), 389–423 (1983)

121. Chen, Y.: A new family of surfaces of general type with K2=7 and pg=0. Math. Z. 275(3–4), 1275–1286 (2013)

122. Chen, X., Donaldson, S., Sun, S.: Kähler-Einstein metrics and stability. Int. Math. Res. Not. IMRN 8, 2119–2125 (2014)

123. Chen, X., Donaldson, S., Sun, S.: Kähler-Einstein metrics on Fano manifolds. I: Approximation of metrics with cone singularities. II: Limits with cone angle less than 2π. and completion of the main proof. J. Amer. Math. Soc. 28(1), 183–197, 199–234, 235–278 (2015)

124. Chevalley, C.: Invariants of finite groups generated by reflections. Am. J. Math. 77, 778–782 (1955)

125. Clebsch, A.: Zur Theorie der Riemann'schen Flächen. Math. Ann. 6, 216–230 (1872)

126. Clemens, H.: Geometry of formal Kuranishi theory. Adv. Math. 198(1), 311–365 (2005)

127. Cornalba, M.: On the locus of curves with automorphisms. Ann. Mat. Pura Appl. 149(4), 135–151 (1987). Erratum: Ann. Mat. Pura Appl. (4) 187(1) 185–186 (2008)

128. Corti, A.: Factoring birational maps of threefolds after Sarkisov. J. Algebraic Geom. 4(2), 223–254 (1995)

129. Costa, A.F., Izquierdo, M., Parlier, H.: Connecting p-gonal loci in the compactification of moduli space.arXiv:1305.0284

130. de Franchis, M.: Sulle superficie algebriche le quali contengono un fascio irrazionale di curve. Palermo Rend. 20, 49–54 (1905)

131. De Concini, C., Salvetti, M.: Stability for the cohomology of Artin groups. Adv. Math. 145(2), 291–305 (1999)

132. Dehn, M.: Die Gruppe der Abbildungsklassen (Das arithmetische Feld auf Flächen). Acta Math. 69, 135–206 (1938)

133. Deligne, P.: Le groupe fondamental du complément d'une courbe plane n'ayant que des points doubles ordinaires est abélien (d'après W. Fulton). [The fundamental group of the complement of a plane curve having only ordinary double points is abelian (after W. Fulton)] Bourbaki Seminar, Vol. 1979/80, pp. 1–10. Lecture Notes in Math., 842. Springer, Berlin-New York (1981)

134. Deligne, P., Griffiths, P., Morgan, J., Sullivan, D.: Real homotopy theory of Kähler manifolds. Invent. Math.29(3), 245–274 (1975)

135. Deligne, P., Mostow, G.D.: Commensurabilities Among Lattices in PU(1, n). Annals of Mathematics Studies, 132. Princeton University Press, Princeton, NJ, pp. viii+183 (1993)

136. Deligne, P., Mumford, D.: The irreducibility of the space of curves of given genus. Inst. Hautes Études Sci. Publ. Math. 36, 75–109 (1969)

137. Dolgachev, I.: Algebraic surfaces with q=pg=0. Algebraic surfaces, 97–215, C.I.M.E. Summer School 1977, 76, Liguori Editore, Napoli (1981), reedited by Springer, Heidelberg (2010)

138. Donaldson, S.K.: Gauge theory and four-manifold topology. [CA] Joseph, A. et al. (ed.) First European congress of mathematics (ECM), Paris, France, July 6–10, 1992. Volume I: Invited lectures (Part 1). Prog. Math. 119. Birkhäuser, Basel, pp. 121–151 (1994)

139. Donaldson, S.K.: The Seiberg-Witten equations and 4-manifold topology. Bull. Am. Math. Soc. (N S)33(1), 45–70 (1996)

140. Drinfel'd, V.G.: On quasitriangular quasi-Hopf algebras and on a group that is closely connected with Gal(¯Q/Q), (Russian). Algebra i Analiz 2(4), 149–181 (1990) [translation in Leningrad Math. J. 2(4), 829–860 (1991)]

141. Dunfield, N.M., Thurston, W.P.: Finite covers of random 3-manifolds. Invent. Math. 166(3), 457–521 (2006)

142. Du Val, P.: On singularities of surfaces which do not impose adjunction conditions. Proc. Camb. Philos. Soc. 30, 483–491 (1934)

143. Easton, R.W., Vakil, R.: Absolute Galois acts faithfully on the components of the moduli space of surfaces: a Belyi-type theorem in higher dimension. Int. Math. Res. Not. IMRN, (20), 10, Art. ID rnm080 (2007) doi:10.1093/imrn/rnm080

144. Eckmann, B., Frölicher, A.: Sur l'intégrabilité des structures presque complexes. C. R. Acad. Sci. Paris232, 2284–2286 (1951)

145. Edmonds, A.L.: Surface symmetry. I. Mich. Math. J. 29(2), 171–183 (1982)

146. Edmonds, A.L.: Surface symmetry. II. Mich. Math. J. 30(2), 143–154 (1983)

147. Eells, J., Sampson, J.H.: Harmonic maps of Riemannian manifolds. Am. J. Math. 86, 109–160 (1964)

148. Ehresmann, C.: Sur les espaces fibrés différentiables. C. R. Acad. Sci. Paris 224, 1611–1612 (1947)

149. Ehresmann, C.: Sur la théorie des espaces fibrés. Topologie algébrique, pp. 3–15. Colloques Internationaux du Centre National de la Recherche Scientifique, no. 12. Centre de la Recherche Scientifique, Paris (1949)

150. Ekedahl, T.: Two examples of smooth projective varieties with nonzero Massey products. Algebra, algebraic topology and their interactions (Stockholm, 1983), 128–132. Lecture Notes in Math., 1183. Springer, Berlin (1986)

151. Ekedahl, T.: Canonical models of surfaces of general type in positive characteristic. Inst. Hautes Études Sci. Publ. Math. 67, 97–144 (1988)

152. Eliashberg, Y., Mishachev, N.: Introduction to the h-principle. Graduate Studies in Mathematics, 48. American Mathematical Society, Providence, RI, pp. xviii+206 (2002)

153. Ellenberg, J.S., Venkatesh, A., Westerland, C.: Homological stability for Hurwitz spaces and the Cohen–Lenstra conjecture over function fields. arXiv:0912.0325

154. Ellia, P.: Codimension two subvarieties and related questions. Vector bundles and low codimensional subvarieties: state of the art and recent developments, 129–208, Quad. Mat., 21, Dept. Math., Seconda Univ. Napoli, Caserta (2007)

155. Enriques, F.: Introduzione alla geometria sopra le superficie algebriche. Memorie della Societa' Italiana delle Scienze (detta "dei XL"), s.3, to. X, 1–81 (1896)

156. Enriques, F., Severi, F.: Mémoire sur les surfaces hyperelliptiques. Acta Math. 32, 283–392 (1909) and 33, 321–403 (1910)

157. Enriques, F.: Le Superficie Algebriche. Zanichelli, Bologna, pp. xv+464 (1949)

158. Erdenberger, C., Grushevsky, S., Hulek, K.: Some intersection numbers of divisors on toroidal compactifications of Ag. J. Algebraic Geom. 19(1), 99–132 (2010)

159. Eyssidieux, P.: Sur la convexité holomorphe des revêtements linéaires réductifs d'une variété projective algébrique complexe. Invent. Math. 156(3), 503–564 (2004)

160. Eyssidieux, P., Katzarkov, L., Pantev, T., Ramachandran, M.: Linear Shafarevich conjecture. Ann. Math. (2) 176(3), 1545–1581 (2012)

161. Eyssidieux, P.: Lectures on the Shafarevich conjecture on uniformization. Complex manifolds, foliations and uniformization, 101–148, Panor. Synthses, 34/35, Soc. Math. France, Paris (2011)

162. Eyssidieux, P., Guedj, V., Zeriahi, A.: Singular Kähler-Einstein metrics. J. Am. Math. Soc. 22(3), 607–639 (2009)

163. Faltings, G., Chai, C.-L.: Degeneration of abelian varieties. With an appendix by David Mumford. Ergebnisse der Mathematik und ihrer Grenzgebiete (3) [Results in Mathematics and Related Areas (3)], 22. Springer-Verlag, Berlin, pp. xii+316 (1990)

164. Fantechi, B.: Stacks for everybody. European Congress of Mathematics, vol. I (Barcelona, 2000), 349–359. Progr. Math., 201. Birkhäuser, Basel (2001)

165. Farb, B., Margalit, D.: A Primer on Mapping Class Groups. Princeton Mathematical Series, vol. 49. Princeton University Press, Princeton, NJ (2012)

166. Freedman, M.: The topology of four-dimensional manifolds. J. Differ. Geom. 17(3), 357–454 (1982)

167. Freedman, M., Quinn, F.: Topology of 4-Manifolds. Princeton mathematical Series, vol. 39. Princeton University Press (1990)

168. Friedman, R., Morgan, J.W.: Algebraic surfaces and four-manifolds: some conjectures and speculations. Bull. Am. Math. Soc. 18, 1–19 (1988)

169. Fried, M.D., Jarden, M.: Field arithmetic, 3rd edn. Revised by Jarden. Ergebnisse der Mathematik und ihrer Grenzgebiete. 3. Folge. A Series of Modern Surveys in Mathematics, 11. Springer-Verlag, Berlin, pp. xxiv+792 (2008)

170. Fried, M.D., Völklein, H.: The inverse Galois problem and rational points on moduli spaces. Math. Ann.290(4), 771–800 (1991)

171. Fulton, W.: Hurwitz schemes and irreducibility of moduli of algebraic curves. Ann. Math. 2(90), 542–575 (1969)

172. Fulton, W.: On the topology of algebraic varieties. Algebraic geometry, Bowdoin, 1985 (Brunswick, Maine, 1985), 15–46. Proc. Sympos. Pure Math., 46, Part 1, Amer. Math. Soc., Providence, RI (1987)

173. Fulton, W.: Equivariant cohomology in algebraic geometry. Lectures at Columbia University (2007)

174. Fulton, W., Hansen, J.: A connectedness theorem for projective varieties, with applications to intersections and singularities of mappings. Ann. Math. (2) 110(1), 159–166 (1979)

175. Fulton, W., Lazarsfeld, R.: Connectivity and its applications in algebraic geometry. Algebraic geometry (Chicago, Ill., 1980), pp. 26–92. Lecture Notes in Math., 862. Springer, Berlin-New York (1981)

176. van der Geer, G.: The cohomology of the moduli space of Abelian varieties. In: Farkas, G., Morrison, I. (eds.) Handbook of Moduli, vol. I, pp. 415–457. Adv. Lect. Math. (ALM), 24. Int. Press, Somerville, MA (2013)

177. Gieseker, D.: Global moduli for surfaces of general type. Invent. Math. 43(3), 233–282 (1977)

178. Gieseker, D.: Lectures on moduli of curves. Tata Institute of Fundamental Research Lectures on Mathematics and Physics, 69. Published for the Tata Institute of Fundamental Research, Bombay. Springer-Verlag, Berlin-New York, pp. iii+99 (1982)

179. Gieseker, D.: Geometric invariant theory and applications to moduli problems. Invariant theory (Montecatini, 1982), 45–73. Lecture Notes in Math., 996. Springer, Berlin (1983)

180. Godement, R.: Topologie algébrique et théorie des faisceaux. Actualités Scientifiques et Industrielles No. 1252. Publ. Math. Univ. Strasbourg. No. 13 Hermann, Paris (1958), pp. viii+283 (reprinted) (1973)

181. Goldman, W.M., Millson, J.J.: The deformation theory of representations of fundamental groups of compact Kähler manifolds. Inst. Hautes Études Sci. Publ. Math. 67, 43–96 (1988)

182. Gompf, R.E.: A new construction of symplectic manifolds. Ann. Math. (2) 142(3), 527–595 (1995)

183. González-Diez, G., Jaikin-Zapirain, A.: The absolute Galois group acts faithfully on regular dessins and on Beauville surfaces (preprint) (2014)

184. Goresky, M., MacPherson, R.: Stratified Morse theory. Ergebnisse der Mathematik und ihrer Grenzgebiete (3) 14. Springer-Verlag, Berlin, pp. xiv+272 (1988)

185. Grauert, H., Remmert, R.: Komplexe Räume. Math. Ann. 136, 245–318 (1958)

186. Grauert, H.: Über die Deformation isolierter Singularitäten analytischer Mengen. Invent. Math. 15, 171–198 (1972)

187. Grauert, H.: Der Satz von Kuranishi für kompakte komplexe Räume. Invent. Math. 25, 107–142 (1974)

188. Green, M.: Restrictions of linear series to hyperplanes, and some results of Macaulay and Gotzmann. Algebraic curves and projective

geometry (Trento, 1988). Lecture Notes in Math., 1389. Springer, Berlin, pp. 76–86 (1989)

189. Greenberg, M.J.: Lectures on Algebraic Topology. W. A. Benjamin Inc, New York-Amsterdam, pp. x+235 (1967). Second edition: Greenberg, M.J., Harper, J.R.: Algebraic topology. A first course. Mathematics Lecture Note Series, 58. Benjamin/Cummings Publishing Co., Inc, Advanced Book Program, Reading, Mass., pp. xi+311 (1981)

190. Green, M., Lazarsfeld, R.: Deformation theory, generic vanishing theorems, and some conjectures of Enriques. Catanese and Beauville. Invent. Math. 90(2), 389–407 (1987)

191. Green, M., Lazarsfeld, R.: Higher obstructions to deforming cohomology groups of line bundles. J. Am. Math. Soc. 4(1), 87–103 (1991)

192. Griffiths, P.: Periods of integrals on algebraic manifolds. I. Construction and properties of the modular varieties. II. Local study of the period mapping. Am. J. Math. 90, 568–626 and 805–865 (1968)

193. Griffiths, P.: Periods of integrals on algebraic manifolds. III. Some global differential-geometric properties of the period mapping. Inst. Hautes Études Sci. Publ. Math. 38, 125–180 (1970)

194. Griffiths, P., Harris, J.: Principles of Algebraic Geometry. Pure and Applied Mathematics. Wiley-Interscience [Wiley], New York, pp. xii+813 (1978)

195. Griffiths, P., Morgan, J.: Rational homotopy theory and differential forms. Progress in Mathematics, 16. Birkhäuser, Boston, Mass., pp. xi+242 (1981)

196. Gromov, M.: Sur le groupe fondamental d'une variété kählérienne. C. R. Acad. Sci. Paris Sér. I 308(3), 67–70 (1989)

197. Gromov, M.: Kähler hyperbolicity and L2-Hodge theory. J. Differ. Geom. 33(1), 263–292 (1991)

198. Gromov, M., Schoen, R.: Harmonic maps into singular spaces and p-adic superrigidity for lattices in groups of rank one. Inst. Hautes Études Sci. Publ. Math. 76, 165–246 (1992)

199. Grothendieck, A.: Sur quelques points d'algèbre homologique. Tohoku Math. J. II. Ser 9, 119–221 (1957)

200. Grothendieck, A.: Techniques de construction et théoremes d'existence en géométrie algébrique. IV, Les schemas de Hilbert, Sem. Bourbaki, 13, 1–28 (1960–61)

201. Grothendieck A.: (dirigé par), Revêtements étales et groupe fondamental, Séminaire de géométrie algébrique du Bois Marie 1960–61 Springer Lecture Notes in Math. 224, (1971). Reedited by Société Mathématique de France in the series 'Documents mathématiques (2003)

202. Grothendieck, A.: Esquisse d'un programme. In: Geometric Galois actions, 1. London Math. Soc. Lecture Note Ser., 242. Cambridge Univ. Press, Cambridge, pp. 5–48 (1997)

203. Grushevsky, S., Hulek, K., Tommasi, O.: Stable cohomology of the perfect cone toroidal compactification of the moduli space of abelian varieties. arXiv:1307.4646

204. Guenancia, H.: Kähler-Einstein metrics with cone singularities on klt pairs. Int. J. Math. 24(5), 1350035, 19 pp. (2013)

205. Hacon, C.D., McKernan, J.: Boundedness of pluricanonical maps of varieties of general type. Invent. Math. 166(1), 1–25 (2006)

206. Hamilton, R.S.: Three-manifolds with positive Ricci curvature. J. Differ. Geom. 17, 255–306 (1982)

207. Hano, J.: On Kaehlerian homogeneous spaces of unimodular Lie groups. Am. J. Math. 79, 885–900 (1957)

208. Harer, J.L.: Stability of the homology of the mapping class groups of orientable surfaces. Ann. Math. (2) 121(2), 215–249 (1985)

209. Hartmann, P.: On homotopic harmonic maps. Can. J. Math. 19, 673–687 (1967)

210. Hartshorne, R.: Varieties of small codimension in projective space. Bull. Am. Math. Soc. 80, 1017–1032 (1974)

211. Hartshorne, R.: Algebraic Geometry. Springer GTM 52, Berlin (1977)

212. Hatcher, A., Thurston, W.: A presentation for the mapping class group of a closed orientable surface. Topology 19(3), 221–237 (1980)

213. Helgason, S.: Differential Geometry, Lie Groups, and Symmetric Spaces. Pure and Applied Mathematics, 80. Academic Press, Inc. [Harcourt Brace Jovanovich, Publishers], New York-London, pp. xv+628 (1978)

214. Hilton, P.J., Stammbach, U.: A Course in Homological Algebra, 2nd edn. Graduate Texts in Mathematics, vol. 4. Springer-Verlag, New York-Berlin, pp. ix+338 (1971)

215. Hopf, H.: Fundamentalgruppe und zweite Bettische Gruppe. Comment. Math. Helv. 14, 257–309 (1942)

216. Horikawa, E.: On deformations of quintic surfaces. Invent. Math. 31, 43–85 (1975)

217. Hubbard, J.H.: Teichmüller theory and applications to geometry, topology, and dynamics, vol. 1. Teichmüller theory. Matrix Editions, Ithaca, NY, pp. xx+459 (2006)

218. Humphreys, J.E.: Linear Algebraic Groups. Graduate Texts in Mathematics 21. Springer-Verlag, New York-Heidelberg-Berlin, XV, p. 247 (1975)

219. Hurewicz, W., Wallman, H.: Dimension Theory. Princeton University Press, Princeton, p. 165 (1948)

220. Hurwitz, A.: Ueber Riemann'schen Flächen mit gegebenen Verzweigungspunkten. Math. Ann. 39, 1–61 (1891)

221. Igusa, J.: Theta Functions. Die Grundlehren der mathematischen Wissenschaften, Band 194. Springer-Verlag, New York-Heidelberg, pp. x+232 (1972)

222. Inoue, M.: Some new surfaces of general type. Tokyo J. Math. 17(2), 295–319 (1994)

223. Jacobson, N.: Basic Algebra. II. W. H. Freeman and Co., San Francisco, Calif., pp. xix+666 (1980)

224. Kas, A., Schlessinger, M.: On the versal deformation of a complex space with an isolated singularity. Math. Ann. 196, 23–29 (1972)

225. Johnson, F.E.A., Rees, E.G.: On the fundamental group of a complex algebraic manifold. Bull. Lond. Math. Soc. 19(5), 463–466 (1987)

226. Jost, J., Yau, S.-T.: Harmonic mappings and Kähler manifolds. Math. Ann. 262, 145–166 (1983)

227. Jost, J., Yau, S.-T.: A strong rigidity theorem for a certain class of compact complex analytic surfaces. Math. Ann. 271, 143–152 (1985)

228. Jost, J., Yau, S.-T.: Applications of quasilinear PDE to algebraic geometry and arithmetic lattices. Algebraic geometry and related topics (Inchon, 1992), 169–193. In: Conf. Proc. Lecture Notes Algebraic Geom., I, Int. Press, Cambridge, MA (1993)

229. Jost, J., Zuo, K.: Harmonic maps and Sl(r,C)-representations of fundamental groups of quasiprojective manifolds. J. Algebraic Geom. 5(1), 77–106 (1996)

230. Jost, J., Zuo, K.: Harmonic maps of infinite energy and rigidity results for representations of fundamental groups of quasiprojective varieties. J. Differ. Geom. 47(3), 469–503 (1997)

231. Jost, J., Zuo, K.: Harmonic maps into Bruhat-Tits buildings and factorizations of p-adically unbounded representations of π1 of algebraic varieties. I. J. Algebraic Geom. 9(1), 1–42 (2000)

232. Open problems: classification of algebraic and analytic manifolds. Classification of algebraic and analytic manifolds. In: Proc. Symp. Katata/Jap. 1982. Edited by Kenji Ueno. Progress in Mathematics, 39. Birkhauser, Boston, Mass., pp. 591–630 (1983)

233. Kato, M.: A non Kähler structure on an S2-bundle over a ruled surface. Unpublished manuscript, p. 15, May (1992)

234. Katzarkov, L., Ramachandran, M.: On the universal coverings of algebraic surfaces. Ann. Sci. École Norm. Sup. (4) 31(4), 525–535 (1998)

235. Kazhdan D.: Arithmetic varieties and their fields of quasi-definition. Actes du Congrès International des Mathematiciens (Nice, 1970), Tome 2, pp. 321–325. Gauthier-Villars, Paris (1971)

236. Kazhdan, D.: On arithmetic varieties II. Israel J. Math. 44(2), 139–159 (1983)

237. Kerckhoff, S.P.: The Nielsen realization problem. Ann. Math. (2) 117(2), 235–265 (1983)

238. Keum, Y.H.: Some new surfaces of general type with pg=0. Unpublished manuscript (1988)

239. Kharlamov, V.M., Kulikov, V.S.: On real structures of real surfaces. Izv. Ross. Akad. Nauk Ser. Mat. 66(1) 133–152 (2002); translation in. Izv. Math. 66(1), 133–150 (2002)

240. Kharlamov, V.M. Kulikov, V.S.,: The semigroups of coverings. Izv. Ross. Akad. Nauk Ser. Mat. 77(3), 163–198 (2013); translation in. Izv. Math. 77(3), 594–626 (2013)

241. Klingler, B.: Symmetric differentials, Kähler groups and ball quotients. Invent. Math. 192(2), 257–286 (2013)

242. Kobayashi, S.: Geometry of bounded domains. Trans. Am. Math. Soc. 92, 267–290 (1959)

243. Kobayashi, S.: First Chern class and holomorphic tensor fields. Nagoya Math. J. 77, 5–11 (1980)

244. Kobayashi, S.: The first Chern class and holomorphic symmetric tensor fields. J. Math. Soc. Jpn. 32(2), 325–329 (1980)

245. Kobayashi, S., Nomizu, K.: Foundations of Differential Geometry, vol I. Interscience Publishers, a division of John Wiley & Sons, New York-London, pp. xi+329 (1963)

246. Kodaira, K., Morrow, J.: Complex Manifolds. Holt, Rinehart and Winston Inc, New York-Montreal, Que.-London, pp. vii+192 (1971)

247. Kodaira, K.: On Kähler varieties of restricted type (an intrinsic characterization of algebraic varieties). Ann. Math. 2(60), 28–48 (1954)

248. Kodaira, K., Spencer, D.C.: On deformations of complex analytic structures. I, II. Ann. Math. 67, 328–466 (1958)

249. Kodaira, K.: On compact complex analytic surfaces. I. Ann. Math. 71, 111–152 (1960)

250. Kodaira, K.: On stability of compact submanifolds of complex manifolds. Am. J. Math. 85, 79–94 (1963)

251. Kodaira, K.: A certain type of irregular algebraic surfaces. J. Anal. Math. 19, 207–215 (1967)

252. Kodaira, K.: Complex Manifolds and Deformation of Complex Structures. Translated from the Japanese by Kazuo Akao. With an appendix by Daisuke Fujiwara. Grundlehren der Mathematischen Wissenschaften, 283. Springer-Verlag, New York, pp. x+465 (1986)

253. Kollár, J., Shepherd Barron, N.I.: Threefolds and deformations of surface singularities. Invent. Math. 91, 299–338 (1988)

254. Kollár, J.: Shafarevich maps and plurigenera of algebraic varieties. Invent. Math. 113(1), 177–215 (1993)

255. Kollár, J. : Shafarevich Maps and Automorphic Forms. M. B. Porter Lectures. Princeton University Press, Princeton, NJ, pp. x+201 (1995)

256. Kollár, J.: Moduli of varieties of general type. Farkas, G., Morrison, I. (eds.) Handbook of moduli, vol. II, pp. 131–157. Adv. Lect. Math. (ALM), 25. Int. Press, Somerville, MA (2013)

257. Kollár, J., Mori, S.: Birational Geometry of Algebraic Varieties. With the collaboration of C. H. Clemens and A. Corti. Cambridge Tracts in Mathematics, 134. Cambridge University Press, Cambridge, pp. viii+254 (1998)

258. Kotschick, D.: On regularly fibered complex surfaces. In: Proceedings of the Kirbyfest (Berkeley, CA, 1998), 291–298 (electronic), Geom. Topol. Monogr., 2, Geom. Topol. Publ., Coventry (1999)

259. Kuranishi, M.: On the locally complete families of complex analytic structures. Ann. Math. 2(75), 536–577 (1962)

260. Kuranishi, M.: New proof for the existence of locally complete families of complex structures. Proceedings of conference on complex analysis, Minneapolis 1964, 142–154 (1965)

261. Kuranishi, M.: A note on families of complex structures. In: Iyanaga, S., Spencer, D.C. (eds.) Global Analysis (Papers in Honor of K. Kodaira), pp. 309–313, Princeton Univ. Press and Univ. Tokyo Press, Tokyo (1969)

262. Kuranishi, M.: Deformations of compact complex manifolds. Séminaire de Mathématiques Supérieures, No. 39 (Été 1969). Les Presses de l'Université de Montréal, Montreal, Que., pp. 99 (1971)

263. Lange, H., Birkenhake, C.: Complex Abelian Varieties. Grundlehren der Mathematischen Wissenschaften, 302. Springer-Verlag, Berlin, pp. viii+435 (1992); 2nd edn, pp. xii+635 (2004)

264. Lange, H.: Hyperelliptic varieties. Tohoku Math. J. (2) 53(4), 491–510 (2001)

265. Lange, H., Recillas, S.: Abelian varieties with group action. J. Reine Angew. Math. 575, 135–155 (2004)

266. Larsen, M.E.: On the topology of complex projective manifolds. Invent. Math. 19, 251–260 (1973)

267. Lazarsfeld, R.: Positivity in Algebraic Geometry. I. Classical Setting: Line Bundles and Linear Series. II. Positivity for vector bundles, and multiplier ideals. Ergebnisse der Mathematik und ihrer Grenzgebiete. 3. Folge, 48 and 49. Springer-Verlag, Berlin, xviii+387 resp. pp. xviii+385 (2004)

268. LeBrun, C.: Twistors for tourists: a pocket guide for algebraic geometers. Algebraic geometry-Santa Cruz 1995, 361–385. In: Proc. Sympos. Pure Math., 62, Part 2, Amer. Math. Soc., Providence, RI (1997)

269. Lefschetz, S.: On certain numerical invariants of algebraic varieties with application to abelian varieties. Trans. Am. Math. Soc. 22(3), 327–406 (1921)

270. Lefschetz, S.: L'analysis situs et la géométrie algébrique. Gauthier-Villars, Paris, pp. vi+154 (1924)

271. Liu, W.: Stable degenerations of surfaces isogenous to a product II. Trans. A.M.S. 364(5), 2411–2427 (2012)

272. Livingston, C.: Stabilizing surface symmetries. Mich. Math. J. 32, 249–255 (1985)

273. Lönne, M., Penegini, M.: On asymptotic bounds for the number of irreducible components of the moduli space of surfaces of general type. arXiv:1402.6438

274. Madsen, I., Weiss, M.: The stable moduli space of Riemann surfaces: Mumford's conjecture. Ann. Math. (2) 165(3), 843–941 (2007)

275. Magaard, K., Shaska, T., Shpectorov, S., Völklein, H.: The locus of curves with prescribed automorphism group. Communications in arithmetic fundamental groups (Kyoto, 1999/2001). Sūrikaisekikenkyūsho Kōkyūroku No. 1267, pp. 112–141 (2002)

276. Magnus, W., Karrass, A., Solitar, D.: Combinatorial Group Theory: Presentations of Groups in Terms of Generators and Relations. Interscience Publishers [John Wiley-Sons, Inc.], New York-London-Sydney (1966), pp. xii+444 (Dover reprint) (1976)

277. Manetti, M.: Iterated double covers and connected components of moduli spaces. Topology 36(3), 745–764 (1997)

278. Manetti, M.: On the moduli space of diffeomorphic algebraic surfaces. Invent. Math. 143, 29–76 (2001)

279. Manetti, M.: Cohomological constraint on deformations of compact Kähler manifolds. Adv. Math. 186(1), 125–142 (2004)

280. Manetti, M.: Differential graded Lie algebras and formal deformation theory. Algebraic geometry–Seattle 2005. Part 2, 785–810. In: Proceedings of symposium on pure mathematics, 80. Part 2, Amer. Math. Soc., Providence, RI (2009)

281. Markovic, V.: Realization of the mapping class group by homeomorphisms. Invent. Math. 168(3), 523–566 (2007)

282. Mather, J.: Notes on Topological Stability. Harvard University, Math. Notes (1970)

283. Matsumura, H.: On algebraic groups of birational transformations. Atti Accad. Naz. Lincei Rend. Cl. Sci. Fis. Mat. Natur. (8) 34, 151–155 (1963)

284. Meersseman, L.: Global moduli space of complex structures, groupoids and analytic stacks. arXiv:1311.4170v1

285. Migliorini, L.: A smooth family of minimal surfaces of general type over a curve of genus at most one is trivial. J. Algebraic Geom. 4(2), 353–361 (1995)

286. Miller, C.: The second homology group of a group; relations among commutators. Proc. Am. Math. Soc.3, 588–595 (1952)

287. Milne, J.S.: Kazhdan's theorem on arithmetic varieties. arXiv:math/0106197

288. Milnor, J.W.: Morse theory. Based on lecture notes by M. Spivak and R. Wells. Annals of Mathematics Studies, No. 51. Princeton University Press, Princeton, NJ, pp. vi+153 (1963)

289. Milnor, J.W.: Singular Points of Complex Hypersurfaces. Annals of Mathematics Studies 61. Princeton University Press and the University of Tokyo Press, p. 122 (1968)

290. Milnor, J.W.: Introduction to Algebraic K-Theory. Annals of Mathematics Studies 72. Princeton University Press, Princeton (1971)

291. Milnor, J.W., Stasheff, J.D.: Characteristic Classes. Annals of Mathematics Studies. No.76. Princeton University Press and University of Tokyo Press, Princeton, NJ, VII, p. 331 (1974)

292. Milnor, J.: Curvatures of left invariant metrics on Lie groups. Adv. Math. 21(3), 293–329 (1976)

293. Moishezon, B.: The arithmetics of braids and a statement of Chisini. In: Geometric Topology, Haifa 1992. Contemp. Math. 164. Amer. Math. Soc., Providence, RI, pp. 151–175 (1994)

294. Mok, N.: The holomorphic or anti-holomorphic character of harmonic maps into irreducible compact quotients of polydiscs. Math. Ann. 272, 197–216 (1985)

295. Mok, N.: Strong rigidity of irreducible quotients of polydiscs of finite volume. Math. Ann. 282, 555–578 (1988)

296. Mok, N.: Characterization of certain holomorphic geodesic cycles on quotients of bounded symmetric domains in terms of tangent subspaces. Compos. Math. 132(3), 289–309 (2002)

297. Moravec, P.: Unramified Brauer groups of finite and infinite groups. Am. J. Math. 134(6), 1679–1704 (2012)

298. Morgan, J.W.: The Seiberg-Witten Equations and Applications to the Topology of Smooth Four-Manifolds. Mathematical Notes 44. Princeton Univ. Press vi, NY (1996)

299. Morgan, J., Tian, G.: Ricci Flow and the Poincaré Conjecture. Clay Mathematics Monographs 3. American Mathematical Society (AMS), Providence, RI; Clay Mathematics Institute, Cambridge, MA, xlii, p. 521 (2007)

300. Mostow, G.D.: Strong Rigidity of Locally Symmetric Spaces. Annals of Mathematics Studies, no. 78. Princeton University Press, Princeton, NJ; University of Tokyo Press, Tokyo (1973)

301. Mumford, D.: The canonical ring of an algebraic surface. Ann. of Math. (2) 76, 612–615, appendix to The theorem of Riemann-Roch for high multiples of an effective divisor on an algebraic surface, by O. Zariski, ibid. 560–611 (1962)

302. Mumford, D.: Geometric invariant theory Ergebnisse der Mathematik und ihrer Grenzgebiete, Neue Folge, Band 34 Springer-Verlag, Berlin-New York, pp. vi+145 (1965)

303. Mumford, D.: Abelian varieties, Tata Institute of Fundamental Research Studies in Mathematics, No. 5. Published for the Tata Institute of Fundamental Research, Bombay; Oxford University Press, London, pp. viii+242 (1970)

304. Mumford, D.: Stability of projective varieties. Enseignement Math. (2) 23(1–2), 39–110 (1977)

305. Mumford, D.: Hirzebruch's proportionality theorem in the noncompact case. Invent. Math. 42, 239–272 (1977)

306. Mumford, D.: Algebraic geometry. I: complex projective varieties. Grundlehren der mathematischen Wissenschaften, 221. Springer-Verlag, Berlin-Heidelberg-New York, X, p. 186 (1976)

307. Mumford, D.: Towards an enumerative geometry of the moduli space of curves. In: Arithmetic and geometry. Pap. dedic. I. R. Shafarevich, vol. II: Geometry. Prog. Math. 36, Birkhäuser, pp. 271–328 (1983)

308. Mumford, D., Suominen, K.: Introduction to the theory of moduli. In: Algebraic geometry, Oslo 1970 (Proc. Fifth Nordic Summer-

School in Math.), pp. 171–222, Wolters-Noordhoff, Groningen (1972)

309. Mumford, D.: Intuition and rigor and Enriques's quest. Notices Am. Math. Soc. 58(2), 250–260 (2011)

310. Naie, D.: Surfaces d'Enriques et une construction de surfaces de type général avec pg=0. Math. Z.215(2), 269–280 (1994)

311. Namikawa, Y.: Toroidal compactification of Siegel spaces. Lecture Notes in Mathematics, 812. Springer, Berlin, pp. viii+162 (1980)

312. Newlander, A., Nirenberg, L.: Complex analytic coordinates in almost complex manifolds. Ann. Math.2(65), 391–404 (1957)

313. Nielsen, J.: Untersuchungen zur Topologie der geschlossenen zweiseitigen Flächen I, II, III. Acta Math. 50(1), 189–358 (1927), 53(1), 1–76 (1929), 58(1), 87167 (1932)

314. Nielsen, J.: Abbildungsklassen endlicher Ordnung. Acta Math. 75, 23–115 (1943)

315. Nielsen, J.: Surface transformation classes of algebraically finite type. Danske Vid. Selsk. Math.-Phys. Medd. 21(2), 89 (1944)

316. Oguiso, K., Sakurai, J.: Jun Calabi-Yau threefolds of quotient type. Asian J. Math. 5(1), 43–77 (2001)

317. Pardini, R.: Abelian covers of algebraic varieties. J. Reine Angew. Math. 417, 191–213 (1991)

318. Penegini, M., Polizzi, F.: On surfaces with pg=q=2,K2=5 and Albanese map of degree 3. Osaka J. Math. 50(3), 643–686 (2013)

319. Penegini, M., Polizzi, F.: Surfaces with pg=q=2,K2=6 and Albanese map of degree 2. Canad. J. Math.65(1), 195–221 (2013)

320. Perelman, G.: The entropy formula for the Ricci flow and its geometric applications. arXiv:math/0211159

321. Perelman, G.: Ricci flow with surgery on three-manifolds. arXiv:math/0303109

322. Perelman, G.: Finite extinction time for the solutions to the Ricci flow on certain three-manifolds. arXiv:math/0307245

323. Peternell, T., Le Potier, J., Schneider, M.: Vanishing theorems, linear and quadratic normality. Invent. Math. 87(3), 573–586 (1987)

324. Pontecorvo, M.: On the complex geometry of twistor spaces. Geometry Seminars, 1994–1995 (Italian) (Bologna), pp. 195–216, Univ. Stud. Bologna, Bologna (1996)

325. Quinn, F.: Isotopy of 4-manifolds. J. Differ. Geom. 24(3), 343–372 (1986)

326. Reid, M.: Young person's guide to canonical singularities. In: Algebraic geometry, Proc. Summer Res. Inst., Brunswick/Maine 1985. Part 1, Proc. Symp. Pure Math. 46, 345–414 (1987)

327. Ries, J.F.X.: Subvarieties of moduli space determined by finite groups acting on surfaces. Trans. Am. Math. Soc. 335(1), 385–406 (1993)

328. Rollenske, S.: Compact moduli for certain Kodaira fibrations. Ann. Sc. Norm. Super. Pisa Cl. Sci. IX(4), 851–874 (2010)

329. Roos, G.: Jordan Triple Systems, pp. 183–282. In: Faraut J., Kaneyuki S., Korányi A., Lu Q.K., Roos G.: Analysis and Geometry on Complex Homogeneous Domains. Progress in Mathematics, vol. 185, Birkhäuser, Boston (2000)

330. Sampson, J.H.: Some properties and applications of harmonic mappings. Ann. Sci. Éc. Norm. Supér. (4) 11(2), 211–228 (1978)

331. Sampson, J.H.: Applications of harmonic maps to Kähler geometry. Contemp. Math. 49, 125–134 (1986)

332. Seidel, P.: Lectures on four-dimensional Dehn twists. In: Catanese, F., Tian, G. (eds.) Symplectic 4-manifolds and algebraic

surfaces, 231–267. Lecture Notes in Math., 1938. Springer, Berlin (2008)

333. Seifert, H., Threlfall, W.: Lehrbuch der Topologie. IV + 353 S. 132 Abb. Leipzig, B. G. Teubner (1934)

334. Sernesi, E.: Deformations of Algebraic Schemes. Grundlehren der Mathematischen Wissenschaften, 334. Springer-Verlag, Berlin, pp. xii+339 (2006)

335. Serre, J.-P.: Sur la topologie des variétés algébriques en caractéristique p., 1958 Symposium internacional de topología algebraica, pp. 24–53, Universidad Nacional Autónoma de México and UNESCO, Mexico City

336. Serre, J.-P.: Exemples de variétés projectives conjuguées non homéomorphes. C. R. Acad. Sci. Paris258, 4194–4196 (1964)

337. Shafarevich, I.R.: Basic algebraic geometry. Translated from the Russian by K. A. Hirsch. Revised printing of Grundlehren der mathematischen Wissenschaften, Vol. 213, 1974. Springer Study Edition. Springer-Verlag, Berlin-New York, pp. xv+439 (1977)

338. Shephard, G.C., Todd, J.A.: Finite unitary reflection groups. Can. J. Math. 6, 274–304 (1954)

339. Siegel, C.L.: Analytic Functions of Several Complex Variables. Notes by P. T. Bateman. Institute for Advanced Study, Princeton, NJ, pp. ii+200 (1950)

340. Siegel, C.L.: Topics in complex function theory, vol. III. Abelian functions and modular functions of several variables. Translated from the German by E. Gottschling and M. Tretkoff. With a preface by Wilhelm Magnus. Reprint of the 1973 original. Wiley Classics Library. A Wiley-Interscience Publication. Wiley, New York, pp. x+244 (1989)

341. Simpson, C.T.: Higgs bundles and local systems. Inst. Hautes Études Sci. Publ. Math. 75, 5–95 (1992)

342. Simpson, C.T.: Subspaces of moduli spaces of rank one local systems. Ann. Sci. École Norm. Sup. (4) 26(3), 361–401 (1993)

343. Singerman, D.: Finitely maximal Fuchsian groups. J. Lond. Math. Soc. 6, 29–38 (1972)

344. Siu, .T.: The complex-analyticity of harmonic maps and the strong rigidity of compact Kähler manifolds. Ann. Math. (2) 112(1), 73–111 (1980)

345. Siu, Y.T.: Strong rigidity of compact quotients of exceptional bounded symmetric domains. Duke Math. J.48(4), 857–871 (1981)

346. Siu, Y.-T.: Complex-analyticity of harmonic maps, vanishing and Lefschetz theorems. J. Differ. Geom. 17, 55–138 (1982)

347. Siu, Y.-T.: Strong rigidity for Kähler manifolds and the construction of bounded holomorphic functions. Discret. Groups Geom. Anal. Pap. Hon. G. D. Mostow 60th Birthday, Prog. Math. 67, 124–151 (1987)

348. Sommese, A.J.: Quaternionic manifolds. Math. Ann. 212, 191–214 (1974/75)

349. Spanier, E.H.: Algebraic Topology. McGraw-Hill Book Co., New York-Toronto, Ont.-London, pp. xiv+528 (1966)

350. Steenrod, N.: The Topology of Fibre Bundles. Princeton Mathematical Series, vol. 14. Princeton University Press, Princeton, NJ, pp. viii+224 (1951)

351. Taubes, C.H.: The existence of anti-self-dual conformal structures. J. Differ. Geom. 36(1), 163–253 (1992)

352. Tian, G., Zhang Z.: Regularity of the Kähler-Ricci flow. C. R. Math. Acad. Sci. Paris 351(15–16), 635–638 (2013); announcement of Regularity of Kähler–Ricci flows on Fano manifolds. arXiv:1310.5897

353. Tjurina, G.N.: Resolution of singularities of plane (=flat) deformations of double rational points. Funk. Anal. i Prilozen 4, 77–83 (1970)

354. Toledo, D.: Projective varieties with non-residually finite fundamental group. Inst. Hautes Études Sci. Publ. Math. 77, 103–119 (1993)

355. Totaro, B.: The Chow ring of a classifying space. In: Algebraic K-theory (Seattle, WA, 1997), 249–281. Proceedings of symposium on pure mathematics, 67. A.M.S., Providence, RI (1999)

356. Tromba, A.: Dirichlet's energy on Teichmüller 's moduli space and the Nielsen realization problem. Math. Z. 222, 451–464 (1996)

357. Uchida, K., Yoshihara, H.: Discontinuous groups of affine transformations of C3. Tohoku Math. J. (2)28(1), 89–94 (1976)

358. Ueno, K.: Classification theory of algebraic varieties and compact complex spaces. Notes written in collaboration with P. Cherenack. Lecture Notes in Mathematics, vol. 439. Springer-Verlag, Berlin-New York, pp. xix+278 (1975)

359. Vakil, R.: Murphy's law in algebraic geometry: badly-behaved deformation spaces. Invent. Math. 164(3), 569–590 (2006)

360. Viehweg, E.: Quasi-projective moduli for polarized manifolds. Ergebnisse der Mathematik und ihrer Grenzgebiete, 3. Folge, 30. Springer-Verlag, Berlin, pp. viii+320 (1995)

361. Viehweg, E.: Compactifications of smooth families and of moduli spaces of polarized manifolds. Ann. Math. 172, 809–910 (2010)

362. Viehweg, E., Zuo, K.: Arakelov inequalities and the uniformization of certain rigid Shimura varieties. J. Differ. Geom. 77(2), 291–352 (2007)

363. Voisin, C.: Théorie de Hodge et géométrie algébrique complexe, Cours Spécialisés, 10, Société Mathématique de France, Paris, pp. viii+595 (2002)

364. Voisin, C.: On the homotopy types of compact Kähler and complex projective manifolds. Invent. Math.157(2), 329–343 (2004)

365. Voisin, C.: On the homotopy types of Kähler manifolds and the birational Kodaira problem. J. Differ. Geom. 72(1), 43–71 (2006)

366. Wajnryb, B.: A simple presentation for the mapping class group of an orientable surface. Israel J. Math.45(2–3), 157–174 (1983)

367. Wajnryb, B.: An elementary approach to the mapping class group of a surface. Geom. Topol. 3, 405–466 (1999)

368. Wall, C.T.C.: On the orthogonal groups of unimodular quadratic forms. Math. Ann. 147, 328–338 (1962)

369. Walker, R.J.: Algebraic curves (Princeton Mathematical Series. 13). Princeton University Press, Princeton, NJ; Oxford University Press, London, X, p. 201 (reprinted also by Dover and Springer) (1950)

370. Washington, L.C.: Introduction to Cyclotomic Fields. Graduate Texts in Mathematics, 83. Springer-Verlag, New York, pp. xi+389 (1982); 2nd edn., pp. xiv+487 (1997)

371. Wavrik, J.: Obstructions to the existence of a space of moduli. Global Analysis (Papers in Honor of K. Kodaira), pp. 403–414. Princeton University and Univ. Tokyo Press, Tokyo (1969)

372. Weibel, C.A.: An Introduction to Homological Algebra. Cambridge Studies in Advanced Mathematics, 38. Cambridge University Press, Cambridge, pp. xiv+450 (1994)

373. Weyl, H.: Invariants. Duke Math. J. 5(3), 489–502 (1939)

374. Wiegold, J.: The Schur multiplier: an elementary approach. Groups-St. Andrews 1981 (St. Andrews, 1981). London Math. Soc. Lecture Note Ser., 71, pp. 137–154, Cambridge Univ. Press (1982)

375. Yau, S.T.: Calabi's conjecture and some new results in algebraic geometry. Proc. Natl. Acad. Sci. USA74, 1798–1799 (1977)

376. Yau, S.T.: On the Ricci curvature of a compact Kähler manifold and the complex Monge-Ampère equation I. Commun. Pure Appl. Math. 31(3), 339–411 (1978)

377. Yau, S.-T.: A splitting theorem and an algebraic geometric characterization of locally Hermitian symmetric spaces. Commun. Anal. Geom. 1(3–4), 473–486 (1993)

378. Zak, F.L.: Projections of algebraic varieties. Mat. Sb. (N.S.) 116(158), 608(4), 593–602 (1981)

379. Zariski, O.: Algebraic Surfaces. With appendices by S. S. Abhyankar, J. Lipman and D. Mumford. Preface to the appendices by Mumford. Ergebnisse der Mathematik und ihrer Grenzgebiete, Band 61, second edition (1971), Springer-Verlag, reprint of the first edition Band III, pp. xii+270 (1935)

380. Zimmermann, B.: Surfaces and the second homology of a group. Monatsh. Math. 104(3), 247–253 (1987)

3

NEGATIVE INDEX JACOBI FORMS AND QUANTUM MODULAR FORMS

Kathrin Bringmann, Thomas Creutzig and Larry Rolen

ABSTRACT

In this paper, we consider the Fourier coefficients of a special class of meromorphic Jacobi forms of negative index considered by Kac and Wakimoto. Much recent work has been done on such coefficients in the case of Jacobi forms of positive index, but almost nothing is known for Jacobi forms of negative index. In this paper we show, from two different perspectives, that their Fourier coefficients have a simple decomposition in terms of partial theta functions. The first perspective uses the language of Lie super algebras, and the second applies the theory of elliptic functions. In particular, we find a new infinite family of rank-crank type partial differential equations generalizing the famous example of Atkin and Garvan. We then describe the modularity properties of these coefficients, showing that they are 'mixed partial theta functions', along the way determining a new class of quantum modular partial theta functions which is of independent interest. In particular, we settle the final cases of a question of Kac concerning modularity properties of Fourier coefficients of certain Jacobi forms.

1. BACKGROUND

Since the introduction of the theory of Jacobi forms by Eichler and Zagier [1], connections between Jacobi forms and modular-type objects have been a question of central interest, with applications to many areas including Siegel modular forms, mock modular forms, and Lie theory. In this paper we study the Fourier coefficients of a special family of negative index Jacobi forms. In particular, consider for $N \in \mathbb{N}, M \in \mathbb{N}_0$ the functions

$$\phi_{M,N}(z;\tau) := \frac{\vartheta\left(z+\frac{1}{2};\tau\right)^M}{\vartheta(z;\tau)^N},$$

where $\vartheta(z;\tau)$ is the usual Jacobi theta function

$$\vartheta(z;\tau) := -i\zeta^{-\frac{1}{2}}q^{\frac{1}{8}}(q)_\infty(\zeta)_\infty\left(\zeta^{-1}q\right)_\infty. \tag{1}$$

Here, $q := e^{2\pi i\tau}(\tau \in \mathbb{H}), \zeta := e^{2\pi iz}(z \in \mathbb{C})$, and for $n \in \mathbb{N}_0 \cup \{\infty\}, (a)_n := (a;q)_n := \prod_{j=0}^{n-1}\left(1-aq^j\right)$ is the q-Pochhammer symbol. The relationship between Jacobi forms and modular forms has appeared in many guises and stems back to important work on holomorphic Jacobi forms, which states that they have theta decompositions relating them to half-integral weight modular forms [1]. The situation for meromorphic positive index Jacobi forms also well understood; a meromorphic Jacobi form of positive index has Fourier coefficients which are*almost mock modular forms*, which in turn are holomorphic parts of *almost harmonic Maass forms*[2]-[6]. Loosely speaking, almost harmonic weak Maass forms are sums of harmonic weak Maass functions under iterates of the raising operator multiplied by almost holomorphic modular forms. In this paper, we describe new decompositions of the Jacobi forms $\phi_{M,N}(z;\tau)$ which complement this long history of previous work on positive index Jacobi forms in the much more mysterious case of negative index. In addition to being of interest in the subject of general Jacobi forms, here we give further applications of such decompositions, focusing on the special subfamily $\phi_N(\tau):=\phi_{0,N}(\tau)$ as they are of great interest in various areas such as number theory, representation theory, combinatorics, and physics. Here we outline just a few such occurrences.

Firstly, for various choices of N, the functions ϕ_N are of combinatorial interest. In particular, the function ϕ_1 is related to the famous Andrews-Dyson-Garvan crank generating function (see (7) and (8)), which was used by Andrews and Garvan to provide a combinatorial explanation for the Ramnaujan congruences for the partition function [7], as postulated by Dyson [8]. In this paper we describe relations between powers of the crank-generating function with certain Appell-Lerch series, giving a new family of partial differential equations (PDEs) generalizing the 'rank-crank PDE' of Atkin and Garvan [9] (see Theorem 3). This beautiful identity of Atkin and Garvan gives a surprising connection between the rank and crank generating functions which can be used to show various congruences-relating ranks and cranks, as well as useful relations between the rank and crank moments [9]. We also note the other examples of similar PDEs related to combinatorics have shown up in, for example Section 3.2 of [10], where the function $\phi_{1,3}(z;\tau)$ is studied in relation to overpartitions.

Secondly, the functions ϕ_N contain information about certain affine vertex algebras and their associated affine Lie algebras studied by Kac and Wakimoto [11]. More precisely, let S(1) be the graded vector space

$$\mathcal{S}(1) := \mathrm{Sym} \bigoplus_{n=0}^{\infty} \left(q^{n+\frac{1}{2}} \left(\zeta \mathbb{C} \oplus \zeta^{-1} \mathbb{C} \right) \right),$$

and $\mathcal{S}(N) = \mathcal{S}(1)^{\otimes N}$. The vector space S(N) can be given the structure of a vertex algebra, the bosonic $\beta\,\gamma$-ghost vertex algebra of rank N and central charge $-N$. The graded character of this vertex algebra has a nice product form in the domain $|q|^{\frac{1}{2}} < |\zeta_n| < |q|^{-\frac{1}{2}}$, where $\zeta_n := e^{2\pi i z_n}$,

$$\mathrm{ch}[\mathcal{S}(N)](z_1, \cdots, z_N; \tau) = \prod_{n=1}^{N} q^{\frac{1}{24}} \mathrm{ch}[\mathcal{S}(1)](z_n; \tau) = q^{\frac{N}{24}} \prod_{n=1}^{N} \frac{1}{\left(\zeta_n^{-1} q^{\frac{1}{2}}\right)_\infty \left(\zeta_n q^{\frac{1}{2}}\right)_\infty}. \tag{2}$$

It specializes to $\phi_N(z;\tau)$ for the following choices:

$$i^N \zeta^{\frac{N}{2}} q^{-\frac{N}{6}} (q)_\infty^{-N} \mathrm{ch}[\mathcal{S}(N)] \left(z - \frac{\tau}{2}, \cdots, z - \frac{\tau}{2}; \tau \right) = \phi_N(z; \tau).$$

The algebra S(N) contains as commuting subalgebras the rank 1 Heisenberg vertex algebra H(1) and the simple affine vertex algebra of $s\ell(N)$ at level −1, $L_{-1}(s\ell(N))$. Note that −1 is not an admissible level in the case of N=2 and that these algebras do not form a mutually commuting pair inside S(2). However, for N>2 it was shown in [12] that $L_{-1}(s\ell(N))$ and H(1) form such a mutually commuting pair inside S(N). The character of the highest-weight module $\mathcal{F}_\mu$, $\mu \in \mathbb{R}$, of $\mathcal{H}(1)$ takes the form

$$\mathrm{ch}\left[\mathcal{F}_\mu(z; \tau)\right] = \frac{\zeta^{\sqrt{N}\mu} q^{-\frac{\mu^2}{2}}}{q^{\frac{1}{24}} (q)_\infty}, \tag{3}$$

so that the Fourier coefficients in ζ of $\phi_N(z;\tau)$ immediately allow one to compute the multiplicity with which the character of $\mathcal{F}_{r/\sqrt{N}}$ appears. In physics language, such a multiplicity is called the *branching function* of the coset S(N)/H(1).

This leads to the second conformal field theory and vertex algebra importance of decomposing a meromorphic Jacobi form. One of the most interesting classes of vertex algebras is given by Vk(g), the universal affine vertex algebra of the simple Lie algebra g at level k∈C. For certain rational admissible levels, Vk(g) is not simple and one instead prefers to study its simple quotient Lk(g). The characters of irreducible highest-weight modules at admissible level Lk(g) are the sum expansions in special domains of meromorphic Jacobi forms [13]. Understanding these sum expansions is crucial in studying the modular data of the corresponding conformal field theory [14],[15].

Fourthly, the functions ϕ_N appear in the denominator identities of affine Lie super algebras [16]. In [14],[15] the denominator identity of $\widehat{s\ell}$^(2|1) was an essential ingredient to study the relations between characters of admissible level $L_k(s\ell(2))$, while we use the identities for the family $\widehat{s\ell}(N|1)$ to prove one of our central theorems.

Finally, the functions ϕ_N also occur in string theory; we only expound upon one example. The reciprocal of the Igusa cusp form $\Phi_{10}(Z)$ (Z∈H2, the Siegel upper half plane of genus (2) arises as the partition function of quarter-BPS dyons in the type II compactification on the product of a K_3 surface and an elliptic curve. Write

$$\frac{1}{\Phi_{10}(Z)} = \sum_{m=-1}^{\infty} \psi_m(z;\tau)\rho^m \qquad \left(Z = \begin{pmatrix} \tau & z \\ z & w \end{pmatrix}, \ \rho = e^{2\pi i w}\right). \tag{4}$$

For $m>0$, the Fourier coefficients of the functions ψ_m are the degeneracies of single-centered black holes and two-centered black holes with total magnetic charge invariant equal to m. This case is studied in pathbreaking work of Dabholkar, Murthy, and Zagier [3] (see also [17] for the appearance of mock modular forms in the context of quantum gravity partition functions and AdS3/CFT2, as well as [18] for a relation between multi-centered black holes and mock Siegel-Narain theta functions). The coefficient of $m=-1$ equals

$$\frac{-1}{\eta^{18}(\tau)\vartheta(z;\tau)^2}$$

(note that our theta function $\vartheta(z;\tau)$ differs from the theta function $\theta_1(z;\tau)$ in the notation of [3] by a factor of i). Analogously to the case of Jacobi forms of positive index, one may view Theorem 3 below as a decomposition giving a 'polar part' but no 'finite part' as described in [2],[3] and stated in more detail in (24). This is consistent with a string theoretic interpretation of ψ_{-1} in (4) in that there are no single-centered black holes and the degeneracies are all interpreted as accounting for two-centered black holes (see [19],[20]). In contrast, in the case $m>0$, the mock part of ψ_m corresponds to single-centered black holes and the Appel Lerch sum corresponds to two-centered black hole bound states [3].

Returning to the problem of studying $\phi_{M,N}$, we define its Fourier coefficients by

$$\phi_{M,N}(z;\tau) =: \sum_{r\in\frac{M-N}{2}+\mathbb{Z}} \chi(M,N,r;\tau)\zeta^r$$

and in particular we set $\chi(N,r;\tau):=\chi(0,N,r;\tau)$. Note that wallcrossing occurs; the coefficients $\chi(M,N,r;\tau)$ are only well-defined if we fix a range for z. We show that the Fourier coefficients $\chi(M,N,r;\tau)$ can be described using *partial theta functions* (i.e., sums over half a lattice which when summed over a full lattices becomes a theta function), whose modularity properties near the real line we also describe using quantum modular forms. Quantum modular forms were recently defined by Zagier in [21] (see also [22]-[24]). Although the definition is not rigorous, Zagier gave a number of motivating examples. Roughly speaking, a *weight k quantum modular form* is a function f:Q→C for some subset Q⊆P1(Q) such that for any γ in a congruence subgroup Γ, the cocycle $f|_k(1-\gamma)$ extends to an open set of and is 'nice' (e.g., continuously differentiable, smooth). In fact, our study of the modularity of the partial theta functions shows that they are what Zagier refers to as strong quantum modular forms, namely, that they have a near-modular property for asymptotic expansions defined at every point in a subset of P1(Q). Moreover, this behavior comes from the 'leaking' of modularity properties of a non-holomorphic Eichler integral defined on the lower half plane (see (38)).

Returning to the Fourier coefficients of $\phi_{M,N}$, we define a *mixed partial theta function* to be a linear combination of quasimodular forms multiplied with partial theta functions. These functions have known connections to many interesting combinatorial functions, such as concave and convex compositions [25], unimodal sequences [26],[27], and stacks [28]. Throughout, we abuse notation to say that any function is a modular form, partial theta function, mixed partial theta function, etc. if it is equal to such a function up to multiplication by a rational power of q. Our main result is the following:

Theorem 1.

For any N∈N, M∈2N0 and r∈Z with $M<N$, the functions $\chi(M,N,r;\tau)$ are mixed partial theta functions.

Remarks.

1) The quasimodular forms appearing in the decomposition of the mixed partial theta functions are canonically determined by the Laurent expansion of the Jacobi form $\phi_{M,N}$ (see Theorem 4).

2) Using the techniques of this paper, it is easy to relax the condition on M to allow any natural number less than N; however, we restrict to even M for notational convenience. Together with Theorem 1 and the works in [2],[3],[5], this settles the final cases of modularity of Kac-Wakimoto characters raised in [11].

We first consider the case of $\phi_N(z;\tau)$, which we study from two perspectives. Our first viewpoint describes the Fourier coefficients as derivatives of partial theta functions of a rescaled version of the root lattice

$$A_{N-1} := \mathbb{Z}\alpha_1 \oplus \cdots \oplus \mathbb{Z}\alpha_{N-1}$$

of $s\ell(N)$. Here, the α_n are the simple roots of $s\ell(N)$, which are linear functionals on the Cartan subalgebra $\mathfrak{h} \cong \mathbb{C}^{N-1}$of $s\ell(N)$. The Gram matrix of A_{N-1} is the Cartan matrix of $s\ell(N)$. We denote the bilinear form by $(|)$ and abbreviate $t^2 := (t|t)$ for t in A_{N-1}. For r in , we define the subset of $\frac{1}{N}A_{N-1}$

$$T_r := \left\{ \sum_{n=1}^{N-1} a_n\alpha_n \middle| a_n = \frac{(N-n)n}{2} - \frac{rn}{N} + (N-1)m_n, m_n \in \mathbb{Z}, \left(m_{N-1} - \frac{1}{2}\right)\left(r - \frac{1}{2}\right) \geq 0 \right\} \tag{5}$$

and its *partial theta function*

$$P_r(\tau) := \sum_{t \in T_r} e^t q^{\frac{t^2}{2(N-1)}}.$$

The e^t are functions on the Cartan subalgebra h defined by $e^t : u \mapsto e^{t(u)}$ Its evaluation for u∈h is then denoted by $P_r(u;\tau)$. We call P_r a partial theta function because the theta function obtained by summing over the complete lattice $\frac{1}{N}A_{N-1}$,

$$\theta_N(\tau) := \sum_{t \in \frac{1}{N}A_{N-1}} e^{t(0)} q^{\frac{t^2}{2(N-1)}},$$

is a modular form of weight $(N-1)/2$ for $\Gamma(M)$ with $M=N^2(N-1)/2$. This statement is true, since θ_N is the theta function of the lattice $\frac{1}{\sqrt{2M}}A_{N-1}$. The level of this lattice is M, and the modularity of theta functions of lattices is discussed for example in [29].

Further, let ∂ be the differential operator

$$\partial_\alpha e^\lambda := (\lambda|\alpha)e^\lambda \quad \text{and} \quad \partial := \prod_{\alpha \in \Delta_0^+} \partial_\alpha \tag{6}$$

where Δ_0^+ is the set of positive roots of $s\ell(N)$. Finally, set $d_N := \prod_{j=1}^N j!$ and let sign$(r)=1$ if $r \geq 0$ and -1 otherwise. Then we have the following.

Theorem 2.

For $N \geq 2$, the r th Fourier coefficient of $\phi_N(z;\tau)$ is given by

$$\begin{aligned}\chi(N,r;\tau) &= \frac{i^N \operatorname{sign}\left(r - \frac{N}{2}\right) q^{\frac{-r^2}{2N}}}{\eta(\tau)^{N(N+1)} d_{N-1}} \sum_{t \in T_{r-\frac{N}{2}}} \prod_{\alpha \in \Delta_0^+} (t|\alpha) q^{\frac{t^2}{2(N-1)}} \\ &= \frac{i^N \operatorname{sign}\left(r - \frac{N}{2}\right) q^{\frac{-r^2}{2N}}}{\eta(\tau)^{N(N+1)} d_{N-1}} \partial P_{r-\frac{N}{2}}(\tau)\Bigg|_{\left\{e^t=1 \middle| t \in \frac{1}{N}A_{N-1}\right\}}.\end{aligned}$$

Remarks.

Using (2) and (3), Theorem 2 implies the character decomposition

$$\mathrm{ch}[\mathcal{S}(N)](z,\cdots,z;\tau)=\sum_{r\in\mathbb{Z}}\mathrm{ch}\left[\mathcal{F}_{\frac{r}{\sqrt{N}}}(z;\tau)\right]\mathrm{ch}[\mathcal{B}_r](\tau),$$

where

$$\mathrm{ch}[\mathcal{B}_r](\tau):=\frac{\mathrm{sign}(r)}{\eta(\tau)^{N^2-1}d_{N-1}}\sum_{t\in T_r}\prod_{\alpha\in\Delta_0^+}(t|\alpha)q^{\frac{t^2}{2(N-1)}}.$$

The ch[Br](τ) are then characters of $L_{-1}(s\ell(N))$.

The proof of the theorem uses the denominator identity of both $s\ell(N|1)$ and $\widehat{s\ell}(N|1)$ as well as Weyl's character formula for $s\ell(N)$.

The case N=1 follows from the denominator identity of $\widehat{g\ell}(1|1)$ (see Example 1). In this case, the Fourier coefficients relate to the characters of a well-known logarithmic conformal field theory, the W(2,3)-algebra of central charge −2. The modularity of the coefficients has been studied from a different perspective in [30].

The second approach is based on a generalization of a deep identity of Atkin and Garvan. To state it, we first recall the rank and crank generating functions (whose combinatorial meanings are not needed in this paper), which arise in many contexts and in particular give combinatorial explanations of Ramnaujan's congruences (for example, see [7],[8],[31]). Specifically, the generating functions are given as follows:

$$\mathcal{R}(\zeta;q):=\frac{(1-z)}{(q)_\infty}\sum_{n\in\mathbb{Z}}\frac{(-1)^n q^{\frac{n}{2}(3n+1)}}{1-\zeta q^n},\qquad \mathcal{C}(\zeta;q):=\frac{(q)_\infty}{(\zeta q)_\infty(\zeta^{-1}q)_\infty}. \tag{7}$$

We also need the normalized versions

$$\mathcal{R}^*(\zeta;q):=\frac{\zeta^{\frac{1}{2}}q^{-\frac{1}{24}}\mathcal{R}(\zeta;q)}{1-\zeta},\qquad \mathcal{C}^*(\zeta;q):=\frac{\zeta^{\frac{1}{2}}q^{-\frac{1}{24}}\mathcal{C}(\zeta;q)}{1-\zeta}.$$

Note that ϕ_N is essentially the N th power of $\mathcal{C}^*$ as for $N\in\mathbb{N}$ we have

$$\phi_N(z;\tau) = i^N \eta(\tau)^{-2N} C^*(\zeta;q)^N. \tag{8}$$

The simplest case of our decomposition relies on the fact that C∗, and thus ϕ_1, is essentially an Appell-Lerch sum thanks to the following classical partial fraction expansion (for example, see Theorem 1.4 of [32]).

$$C^*(\zeta;q) = \frac{\zeta^{\frac{1}{2}}}{\eta(\tau)} \sum_{n\in\mathbb{Z}} \frac{(-1)^n q^{\frac{n(n+1)}{2}}}{1-\zeta q^n}. \tag{9}$$

For the cube of the crank generating function, Atkin and Garvan [9] proved the following rank-crank PDE which is very useful in establishing congruences and relations between the moments of the rank and crank generating functions:

$$2\eta(\tau)^2 C^*(\zeta;q)^3 = \left(6\mathcal{D}_q + \mathcal{D}_\zeta^2\right) \mathcal{R}^*(\zeta;q). \tag{10}$$

Here and throughout $\mathcal{D}_x := x\frac{\partial}{\partial x}$. Note that this gives a description of ϕ_3 in terms of Appell-Lerch sums by (8).

Zwegers [32] nicely generalized (10) for arbitrary odd powers of the crank generating function using the theory of elliptic forms. For similar results using another clever proof, see also the paper of Chan, Dixit, and Garvan [33].

In this paper, we prove a new family of analogous PDEs which are of independent interest. Moreover, we package Zwegers' family of PDEs in a way which illuminates their structure coming from negative index Jacobi forms. To describe this, we need the Appell-Lerch sums

$$F_N(z,u;\tau) := \zeta^{\frac{N}{2}} w^{\frac{N}{2}} \sum_{n\in\mathbb{Z}} \frac{(-w)^{Nn} q^{\frac{N}{2}n(n+1)}}{1-\zeta w q^n}, \tag{11}$$

where $w := e^{2\pi i u}$. We note that these Appell-Lerch sums are similar to the functions $f_z(u;\tau)$ considered in Chapter 3 of [6], which transform as a Jacobi form in u and as a 'mock Jacobi form' in z. We also require the Laurent coefficients of $\phi_{M,N}(z;\tau)$ at $z = 0$:

$$\phi_{M,N}(z;\tau) = \frac{D_N(\tau)}{(2\pi iz)^N} + \frac{D_{N-2}(\tau)}{(2\pi iz)^{N-2}} + \cdots + O(1). \tag{12}$$

Note that only even or odd Laurent coefficients occur, depending on the parity of N, since $\vartheta(-z;\tau)=-\vartheta(z;\tau)$. It is not hard to see that the coefficients D_j are quasimodular forms. Explicitly, they can be computed quickly in terms of the usual Eisenstein series

$$G_k(\tau) := -\frac{B_k}{2k} + \sum_{n\geq 1} \sigma_{k-1}(n)q^n,$$

where $\sigma_k(n) := \sum_{d|n} d^k$ and B_k is the usual k th Bernoulli number. Specifically, it easily follows from the Jacobi triple product formula that

$$\vartheta(z;\tau) = -2\pi z\eta^3(\tau)\, exp\left(-2\sum_{k\geq 1} G_{2k}(\tau)\frac{(2\pi iz)^{2k}}{(2k)!}\right).$$

The following result puts Zwegers family of PDEs as well as our new family of PDEs into a common framework. Setting

$$\delta_e := \begin{cases} 0 \text{ if } N \in 2\mathbb{N}-1, \\ 1 \text{ if } N \in 2\mathbb{N}, \end{cases}$$

we find:

Theorem 3.

For any N∈N, M∈N0, we have

$$\phi_{2M,N+2M}(z;\tau) = (-1)^{1+\delta_e} \sum_{j=0}^{\frac{N-1-\delta_e}{2}+M} \frac{D_{2j+\delta_e+1}(\tau)}{(2j+\delta_e)!} \mathcal{D}_w^{2j+\delta_e}(F_N(z,u;\tau))\big|_{w=1}.$$

Remarks.

Note that Theorem 3 is more explicit than Zwegers' rank-crank type PDEs as it gives the modular coefficients of the PDEs directly from the structure of the Jacobi form $\phi_{M,N}$. Chan, Dixit, and Garvan also remarked that it would be interesting to find such an explicit expression for the quasimodular forms in the decomposition in that case.

It would be interesting to find a Lie theoretic interpretation of the decomposition in Theorem 3.

Armed with the decomposition in Theorem 3 into Appell-Lerch sums, we can easily pick off the Fourier coefficients$\chi(M,N,r;\tau)$ and write them in terms of the Laurent coefficients of $\phi_{M,N}$ and certain partial theta functions

$$\Theta_{\frac{1}{2}+\delta_e}(N,r;\tau) := \sum_{n\geq 0}(-1)^{Nn}\left(n+\frac{r}{N}\right)^{\delta_e} q^{\frac{N}{2}\left(n+\frac{r}{N}\right)^2}.$$

Specifically, if we let

$$\rho(r) := \begin{cases} r & \text{if } r \geq \frac{N}{2}, \\ N-r & \text{if } r < \frac{N}{2}, \end{cases}$$

then the Fourier coefficients of $\phi_{M,N}$ are as follows.

Theorem 4.

For any $N \in \mathbb{N}, M \in \mathbb{N}_0, r \in \frac{N}{2}+\mathbb{Z}, 0 \leq \mathrm{Im}(z) < \mathrm{Im}(\tau),$ we have

$$\chi(2M, N+2M, r;\tau) = (-1)^{1+\delta_e} q^{-\frac{r^2}{2N}} \sum_{j=0}^{\frac{N-1-\delta_e}{2}+M} \frac{D_{2j+\delta_e+1}(\tau)}{(2j+\delta_e)!} N^{j+\delta_e} 2^j \mathcal{D}_q^j\left(\Theta_{\frac{1}{2}+\delta_e}(N,\rho(r);\tau)\right).$$

If N>1 is odd, these partial theta functions fit into the pioneering work of Folsom, Ono, and Rhoades [23] which gives startling relations between the asymptotic expansions of the rank and crank generating functions, generalizing and proving beautiful formulas of Ramanujan. Their work shows that $\Theta_{\frac{1}{2}}(N,r;\tau)$ is a strong quantum modular form for odd N>1. Although their theorem does not directly apply for N=1, in this case we essentially obtain an eta quotient which is trivially a quantum modular form at cusps where it vanishes.

For even N, both the hypergeometric representations used to determine quantum sets and the proof of quantum modularity are not applicable. Here we use the innovative approach of Lawrence and Zagier [34] to study quantum modularity properties (see also [35]). A key ingredient in our investigation is a beautiful identity of Warnaar [36] which relates certain partial and false theta functions (see (35)). Our main result for studying quantum modularity for even N is the following, which gives a new family of quantum modular forms.

Theorem 5.

For any N∈2N and $r \in \mathbb{Z}, \Theta_{\frac{3}{2}}(N, r; \tau)$ is a strong quantum modular form with quantum set $\widehat{Q}_{N,r}$ (defined in (34)) on $\Gamma_1(2N)$, multiplier system χ_r (defined in (23)), and weight 32.

Remarks.

More details about the specific quantum modular properties can be found in the proof of Theorem 5 in Section 'Quantum modularity of $\Theta_{\frac{3}{2}}(N, r; \tau)$.'

More generally, using Proposition 3 of [37], our proof of Theorem 5 shows that $\Theta_{\frac{3}{2}}(N, r; \tau)$ has modularity properties on all of . For this, we note that although the function is not defined on all of , it has a well-defined asymptotic expansion at all points in . This expansion still agrees with the non-holomorphic Eichler integral on the lower half plane (see Section 'Proof of Theorem 5'), so one could say that $\Theta_{\frac{3}{2}}(N, r; \tau)$ is a quantum modular form on if we allow 'poles' at certain points in .

The paper is organized as follows. In 'Preliminaries on Lie super algebras and character identities' and 'Basic facts on Jacobi forms and quantum modular forms' sections, we review the necessary notation and basic objects from Lie theory, Jacobi forms, and quantum modular forms. We give our first proof of the decomposition using Lie theory in 'The Fourier coefficients and partial theta functions of A_N' section and our second proof using an analogue of the rank-crank PDE in 'Second viewpoint on the decomposition into partial theta functions' section. We conclude by describing the quantum modular properties of $\Theta_{\frac{1}{2}+\nu}(N, r; \tau)$ in 'Proof of Theorem 5' section.

2. PRELIMINARIES ON LIE SUPER ALGEBRAS AND CHARACTER IDENTITIES

In this section, we recall some known facts of the affine Lie superalgebra $\widehat{s\ell}(N|1)$, following [16], as well as the finite-dimensional Lie algebra $s\ell(N)$ using [38].

The Lie super algebra $s\ell(N+1|1)$

In this subsection, the Lie super algebra $s\ell(N+1|1)$ and its root system are defined.

The even subalgebra of the Lie super algebra $s\ell(N+1|1)$ is $g\ell(N+1)$ and the odd part decomposes into the standard representation of the even subalgebra and its conjugate. In order to define the Lie super algebra, it is convenient to first introduce its root system. It lies in the lattice

$$L_N := \mathbb{Z}\varepsilon_1 \oplus \cdots \oplus \mathbb{Z}\varepsilon_{N+1} \oplus \mathbb{Z}\delta$$

with bilinear form

$$\left(\varepsilon_j|\varepsilon_k\right) := \delta_{j,k}, \quad \left(\varepsilon_j|\delta\right) := 0, \quad (\delta|\delta) := -1.$$

Thus, its signature is $(N+1,1)$. The set of roots is $\Delta = \Delta_0 \cup \Delta_1 \subset L_N$, where the set of even roots (respective odd roots) is denoted by Δ_0 (respectively Δ_1). They are

$$\Delta_0 := \left\{\varepsilon_j - \varepsilon_k | 1 \leq j, k \leq N+1, j \neq k\right\}, \qquad \Delta_1 := \left\{\varepsilon_j - \delta, \delta - \varepsilon_j | 1 \leq j \leq N+1\right\}. \tag{13}$$

It is useful to split these sets into positive and negative subroot spaces, where

$$\Delta_0^+ := \left\{\varepsilon_j - \varepsilon_k | 1 \leq j < k \leq N+1\right\}, \quad \Delta_1^+ := \left\{\varepsilon_j - \delta | 1 \leq j \leq N+1\right\},$$
$$\Delta^+ := \Delta_0^+ \cup \Delta_1^+, \qquad \Delta := \Delta^+ \cup \Delta^-.$$

A distinguished system of simple positive roots is then chosen to be

$$\Pi := \left\{\alpha_j = \varepsilon_j - \varepsilon_{j+1}, \beta = \varepsilon_{N+1} - \delta | 1 \leq j \leq N\right\}.$$

The α_j are even roots and β is the only distinguished odd simple root. The inner products of simple positive roots are

$$(\alpha_j|\alpha_k) = \begin{cases} 2 & \text{if } j = k, \\ -1 & \text{if } j = k \pm 1, \\ 0 & \text{otherwise,} \end{cases} \qquad (\beta|\alpha_j) = -\delta_{j,N}, \qquad (\beta|\beta) = 0.$$

Hence, β is an isotropic root. Simple even roots generate the even root lattice

$$A_N := \mathbb{Z}\alpha_1 \oplus \cdots \oplus \mathbb{Z}\alpha_N.$$

Its dual lattice is

$$A'_N := \mathbb{Z}\lambda_1 \oplus \cdots \oplus \mathbb{Z}\lambda_N,$$

where the inner product of the fundamental weights λ_j with simple roots is $(\lambda_j|\alpha_k) = \delta_{j,k}$ and $(\lambda_j|\delta) = 0$. Roots and weights act on the Cartan subalgebra, which is

$$\mathfrak{h} := \bigoplus_{\alpha\in\Pi} \mathbb{C}h_\alpha = \mathfrak{h}_0 \oplus \mathbb{C}h_\beta, \quad \mathfrak{h}_0 := \mathbb{C}h_{\alpha_1} \oplus \cdots \oplus \mathbb{C}h_{\alpha_N},$$

with basis $\{h_\alpha\}$ parameterized by simple positive roots, and h0 the Cartan subalgebra of $s\ell(N+1)$. The fundamental weights λ_j are identified with elements of the dual h∗0 of h0 via λj(hαk)=δj,k. A bilinear form (,) on h is induced from the form on its dual space via

$$(h_\alpha, h_{\alpha'}) := (\alpha|\alpha').$$

We remark that the Lie superalgebra $s\ell(N+1|1)$ is then the Z/2Z-graded algebra generated by $\{h_\alpha, e_\alpha^\pm|\alpha \in \Pi\}$ subject to the Serre-Chevalley relations (14) and the graded Jacobi identity. The parity of h_α and $e_{\alpha_j}^\pm$ is even, while the $e_\beta^\pm$ are odd. We denote the graded anti-symmetric bracket by [,]: $s\ell(N+1|1)\times s\ell(N+1|1)\to s\ell(N+1|1)$. Then the Serre-Chevalley relations of the algebra are

$$[h_\alpha, h_{\alpha'}] = 0, \quad [h_\alpha, e_{\alpha'}^\pm] = \pm(\alpha|\alpha')\, e_{\alpha'}^\pm, \quad [e_\alpha^+, e_{\alpha'}^-] = \delta_{\alpha,\alpha'} h_\alpha, \quad (\mathrm{ad}\, e_\alpha^\pm)^{1-(\alpha|\alpha')}\, e_{\alpha'}^\pm = 0 \tag{14}$$

for all $\alpha, \alpha' \in \Pi$ and $\alpha \neq \alpha'$ in the last equation. The bilinear form (,) on h can be extended to an invariant non-degenerate graded symmetric form on $s\ell(N+1|1)$, which we also denote by (,).

3. THE EVEN WEYL GROUP AND DENOMINATOR IDENTITY OF Sℓ(N+1|1)

We now introduce the even Weyl group and the denominator identity of the Lie super algebra sℓ(N+1|1).

For this, we first need to define the Weyl vector ρ. It is the difference of the even Weyl vector ρ_0 and the odd one ρ_1, namely,

$$\rho_0 := \frac{1}{2}\sum_{\alpha\in\Delta_0^+}\alpha = \frac{1}{2}\sum_{j=1}^{N+1}(N+2-2j)\varepsilon_j, \qquad \rho_1 := \frac{1}{2}\sum_{\alpha\in\Delta_1^+}\alpha = \frac{1}{2}\left(-(N+1)\delta + \sum_{j=1}^{N+1}\varepsilon_j\right),$$

$$\rho := \rho_0 - \rho_1 = \frac{1}{2}\left((N+1)\delta + \sum_{j=1}^{N+1}(N+1-2j)\varepsilon_j\right).$$

The group of even Weyl reflections $W^{\#}$acts on the dual of the even root lattice, A N', and is generated by σ_j,j=1,⋯,N defined by

$$\sigma_j\colon A'_N \to A'_N, \qquad \sigma_j(\lambda) := \lambda - (\alpha_j|\lambda)\,\alpha_j.$$

This action naturally extends to the lattice L_N via $\sigma_j(\delta)$=0 and

$$\sigma_j(\varepsilon_k) := \varepsilon_k - (\varepsilon_k|\alpha_j)\,\alpha_j = \varepsilon_k - (\delta_{j,k} - \delta_{j+1,k})(\varepsilon_j - \varepsilon_{j+1}) = \begin{cases} \varepsilon_{j+1} & j = k, \\ \varepsilon_j & j = k-1, \\ \varepsilon_k & \text{otherwise.} \end{cases}$$

Hence, the even Weyl group $W^{\#}$is just the group S_{N+1} permuting the ε_j. Orthonormality of the ε_j implies that the even Weyl group preserves the bilinear form (|). Following [16] we define

Definition.

A *regular exponential function* on A N' is a finite linear combination of exponentials of the form e^λ for $\lambda \in A'_N$. A*rational exponential function* is the quotient A/B of two regular exponential functions A and B≠0. The even Weyl group $W^{\#}$acts on the field of these functions as $e^\lambda \mapsto e^{w(\lambda)}$ for any w∈$W^{\#}$. The *Weyl denominator* of sℓ(N+1|1) is the rational exponential function

$$R = \frac{\prod_{\alpha \in \Delta_0^+} (1 - e^{-\alpha})}{\prod_{\alpha \in \Delta_1^+} (1 + e^{-\alpha})}.$$

We saw that the even Weyl group $W^{\#}$ is just S_{N+1}, the signum of an element w in $W^{\#}$ is $\sigma(w) := (-1)^n$ if w can be written as a composition of n transpositions. Theorem 2.1 of [16] applied to our situation gives

Lemma 1.

The Weyl denominator of the Lie super algebra $s\ell(N+1|1)$ is

$$e^{\rho} R = \sum_{w \in W^{\sharp}} \sigma(w) w \left(\frac{e^{\rho}}{(1 - e^{\delta - \epsilon_{N+1}})} \right).$$

The denominator identity of the affine Lie super algebra $\widehat{s\ell}(N+1|1)$

We turn our focus to the affinization of $s\ell(N+1|1)$ that is

$$\widehat{s\ell}(N+1|1) := \mathbb{C}\left[t, t^{-1}\right] \otimes s\ell(N+1|1) \oplus \mathbb{C}C \oplus \mathbb{C}d$$

with bracket

$$\left[t^n \otimes x, t^m \otimes y\right] := t^{n+m} \otimes [x, y] + n\delta_{n+m,0}(x, y)\, C, \;\; [d, t^n \otimes x] := nt^n \otimes x, \;\; [C, t^n \otimes x] := [C, d] := 0$$

for all $x,y \in s\ell(N+1|1)$ and $n,m \in Z$. The Cartan subalgebra extends to its affine counterpart

$$\hat{\mathfrak{h}} := \mathfrak{h} \oplus \mathbb{C}d \oplus \mathbb{C}C$$

and the bilinear form extends as

$$\left(t^n \otimes x, t^m \otimes y\right) := (x, y)\delta_{n+m,0}, \;\; (C, d) := 1, \;\; (C, C) := (d, d) := \left(C, t^n \otimes x\right) := \left(d, t^n \otimes x\right) := 0.$$

We identify C and d with linear functionals on h using the bilinear form (,) and extend A_N' to

$$\widehat{A}'_N := A'_N \oplus \mathbb{Z}d \oplus \mathbb{Z}C.$$

The bilinear form extends as

$$(C|d) := 1 \quad \text{and} \quad (C|C) := (d|d) = (C|\lambda) = (d|\lambda) = 0 \quad \forall\lambda \in A'_N.$$

The lattice $A_N \subset A'_N$ is then also a sublattice of $\widehat{A}'_N$. The affine Weyl vector is

$$\widehat{\rho} := \rho + Nd.$$

Note that N is the dual Coxeter number of $s\ell(N+1|1)$. For $\alpha \in A_N$, we define

$$t_\alpha : \widehat{A}'_N \to \widehat{A}'_N, \qquad \lambda \mapsto \lambda + (\lambda|C)\alpha - \left((\lambda|\alpha) + \frac{1}{2}(\alpha|\alpha)(\lambda|C)\right)C. \tag{15}$$

The group of even Weyl translations is $\{t_\alpha | \alpha \in A_N\}$. Conjugation by a Weyl rotation gives for any $w \in W^\natural, \alpha \in A_N$

$$w \circ t_\alpha \circ w^{-1} = t_{w(\alpha)}. \tag{16}$$

Let

$$Y := \left\{h \in \widehat{\mathfrak{h}} | \mathrm{Re}(C(h)) > 0\right\}$$

be the domain of all elements in $\widehat{\mathfrak{h}}$ on which the action of C has positive real part. Let $\widehat{\mathcal{F}}$ be the field of meromorphic functions on Y and define $\mathfrak{q} := e^{-C}$. Thus, $|\mathfrak{q}(y)|<1$ for all y in Y. Any element λ of L' extends to a linear function on $\widehat{\mathfrak{h}}^*$ by defining $\lambda(C)=\lambda(d)=0$. In this way rational exponential functions on L' embed in $\widehat{\mathcal{F}}$.

Definition.

The *denominator* of $\widehat{s\ell}(N+1|1)$ is

$$\widehat{R} = R\prod_{j=1}^{\infty}(1-\mathfrak{q}^j)^{n+1}\prod_{\alpha\in\Delta_0}(1-\mathfrak{q}^j e^\alpha)\prod_{\alpha\in\Delta_1}(1+\mathfrak{q}^j e^\alpha)^{-1}.$$

We need Theorem 4.1 of [16] which states

Lemma 2.

The denominator of $\widehat{s\ell}(N+1|1)$ satisfies

$$e^{\hat{\rho}}\hat{R} = \sum_{\alpha \in A_N} t_\alpha \left(e^{\hat{\rho}} R\right).$$

The Weyl character formula of sℓ(N+1)

We also require a well-known variant of the dimension formula, which itself is a corollary of the famous character formula of Weyl [39]. Let $\lambda = m_1\lambda_1 + \cdots + m_N\lambda_N$ be a dominant weight of sℓ(N+1); that is, all m_j are natural numbers. Letting V_λ be the corresponding irreducible highest-weight module, then the character formula is in our notation

$$\mathrm{ch}[V_\lambda] = \frac{\sum_{w \in W^\sharp} \sigma(w) e^{w(\lambda+\rho_0)}}{e^{\rho_0} \prod_{\alpha \in \Delta_0^+} (1 - e^{-\alpha})}. \qquad (17)$$

Since ρ_1 is $W^\sharp$ invariant, we can replace ρ_0 by ρ in this formula. Let m∈N and let v be the linear map from the regular exponential functions on $\frac{1}{m}A'_N$ to the complex numbers defined by $v(e^\lambda)=1$ for every $\lambda \in \frac{1}{m}A'_N$. Let $V_\lambda$be the irreducible finite-dimensional highest-weight module of highest-weight λ. Hence, v(ch[V_λ]) is just the dimension of this module. The application of v to both nominator and denominator of the character formula (17) vanishes, but the quotient is finite. Using (6), we find [38]

$$v(\mathrm{ch}[V_\lambda]) = \frac{v\left(\partial \sum_{w \in W^\sharp} \sigma(w) e^{w(\lambda+\rho_0)}\right)}{v\left(\partial\, e^{\rho_0} \prod_{\alpha \in \Delta_0^+} (1 - e^{-\alpha})\right)} = \frac{\prod_{\alpha \in \Delta_0^+} (\lambda + \rho_0 | \alpha)}{\prod_{\alpha \in \Delta_0^+} (\rho_0 | \alpha)}.$$

Note that this is Weyl's character formula for irreducible finite-dimensional highest-weight modules. The second equality also holds if we replace $\lambda+\rho_0$ by $z\, w(\lambda+\rho_0)$ for any complex number z and any w in $W^\#$.

Definition.

If m∈N and μ in $\frac{1}{m}A'_N$, then v_μ is the rational exponential function

$$v_\mu = \frac{\sum_{w\in W^\sharp} \sigma(w)e^{w(\mu)}}{e^{\rho_0} \prod_{\alpha\in\Delta_0^+} (1-e^{-\alpha})}. \tag{18}$$

Note that if $\mu-\rho_0$ is dominant, then this is just the character of the irreducible highest-weight module of highest-weight $\mu-\rho_0$. We now closely follow the argument of the proof of the dimension formula of [38].

Lemma 3.

If m∈N and μ in $\frac{1}{m}A'_N$, then

$$\nu(v_\mu) = \frac{1}{d_N}\prod_{\alpha\in\Delta_0^+}(\mu|\alpha) = \frac{\nu(\partial e^\mu)}{d_N}.$$

Proof.

Using the explicit description of the positive even roots in (13), it is easy to compute

$$\prod_{\alpha\in\Delta_0^+}(\rho_0|\alpha) = \prod_{n=1}^{N} n! = d_N.$$

For an arbitrary weight $\mu\in A'_N$, there exists a unique $w\in W^\sharp$ such that $w(\mu+\rho_0)-\rho_0$ is dominant. Letting $\ell(w)$ be the number of positive roots that are mapped to negative ones by w, then $(-1)^{\ell(w)} = \sigma(w)$ (see [38]). Then using that the even Weyl group respects the bilinear form.

$$\nu\left(\nu_{\frac{\mu}{M}}\right)=\nu\left(\frac{\sum\limits_{w\in W^{\natural}}\sigma(w)e^{w\left(\frac{\mu}{M}\right)}}{e^{\rho_0}\prod\limits_{\alpha\in\Delta_0^+}(1-e^{-\alpha})}\right)=(-1)^{\ell(w)}\frac{\prod\limits_{\alpha\in\Delta_0^+}\left(w\left(\frac{\mu}{M}\right)\middle|\alpha\right)}{\prod\limits_{\alpha\in\Delta_0^+}(\rho_0|\alpha)}=\frac{(-1)^{\ell(w)}}{d_N}\prod_{\alpha\in\Delta_0^+}\left(\frac{\mu}{M}\middle|w^{-1}(\alpha)\right)$$

$$=\frac{(-1)^{\ell(w)+\ell(w^{-1})}}{d_N}\prod_{\alpha\in\Delta_0^+}\left(\frac{\mu}{M}\middle|\alpha\right)=\frac{1}{d_N}\prod_{\alpha\in\Delta_0^+}\left(\frac{\mu}{M}\middle|\alpha\right)=\frac{\nu\left(\partial e^{\frac{\mu}{M}}\right)}{d_N}.$$

4. BASIC FACTS ON JACOBI FORMS AND QUANTUM MODULAR FORMS

4.1 Jacobi forms

Here we recall some special Jacobi forms and previous work on Fourier coefficients of Jacobi forms. Jacobi forms are functions from C×H→C which satisfy both an elliptic and a modular transformation law. For the precise definition and basic facts on Jacobi forms, we refer the reader to [1]. In this paper, we are particularly interested in the classical Jacobi theta function, defined in (1). The following transformation laws are well known .

Lemma 4.

For $\lambda,\mu\in\mathbb{Z}$ *and* $\gamma=\left(\begin{smallmatrix}a&b\\c&d\end{smallmatrix}\right)\in\mathrm{SL}_2(\mathbb{Z})$, we have that

$$\vartheta\left(\frac{z}{c\tau+d};\frac{a\tau+b}{c\tau+d}\right)=\psi^3(\gamma)(c\tau+d)^{\frac{1}{2}}e^{\frac{\pi icz^2}{c\tau+d}}\vartheta(z;\tau),$$

$$\vartheta(z+\lambda\tau+\mu;\tau)=(-1)^{\lambda+\mu}q^{-\frac{\lambda^2}{2}}e^{-2\pi i\lambda z}\vartheta(z;\tau),$$

where $\psi(\gamma)$ is the multiplier arising in the transformation law of Dedekind's eta function.

We also require the following theta functions of weight $\frac{1}{2}+\nu$ defined for r∈Z, $\nu\in\{0,1\}$

$$\vartheta_{\frac{1}{2}+\nu}(N,r;\tau):=\sum_{n\in\mathbb{Z}}(-1)^{nN}\left(n+\frac{r}{N}-\frac{1}{2}\right)^{\nu}q^{\frac{N}{2}\left(n+\frac{r}{N}-\frac{1}{2}\right)^2}.$$

We define for convenience the following shifted versions when N∈2N

$$\widetilde{\vartheta}_{\frac{1}{2}+\nu}(N,r;\tau) := \vartheta_{\frac{1}{2}+\nu}\left(N, r+\frac{N}{2};\tau\right). \tag{19}$$

It is trivial to show the following identities:

$$\widetilde{\vartheta}_{\frac{1}{2}+\nu}(N,r+N;\tau) = \widetilde{\vartheta}_{\frac{1}{2}+\nu}(N,r;\tau), \tag{20}$$

$$\widetilde{\vartheta}_{\frac{1}{2}+\nu}(N,-r;\tau) = (-1)^{\nu}\widetilde{\vartheta}_{\frac{1}{2}+\nu}(N,r;\tau), \tag{21}$$

$$\vartheta_{\frac{1}{2}}(1,0;\tau) = \widetilde{\vartheta}_{\frac{3}{2}}(2,r;\tau) = 0. \tag{22}$$

Proposition 1.

If N∈2N and r∈Z, then we have:

$$\widetilde{\vartheta}_{\frac{3}{2}}(N,r;\tau+1) = e\left(\frac{r^2}{2N}\right)\widetilde{\vartheta}_{\frac{3}{2}}(N,r;\tau),$$

$$\widetilde{\vartheta}_{\frac{3}{2}}\left(N,r;-\frac{1}{\tau}\right) = \frac{2}{\sqrt{N}}(-i\tau)^{\frac{3}{2}}\sum_{k=1}^{\frac{N}{2}-1}\sin\left(\frac{2\pi kr}{N}\right)\widetilde{\vartheta}_{\frac{3}{2}}(N,k;\tau).$$

Moreover, for $\nu \in \{0,1\}$, $\widetilde{\vartheta}_{\frac{1}{2}+\nu}(N,r;\tau)$ is a modular form of weight $\frac{1}{2}+\nu$ *on* $\Gamma_1(2N)$ with multiplier

$$\chi_r\begin{pmatrix} a & b \\ c & d \end{pmatrix} := \begin{cases} e\left(\frac{r^2}{2N}\right) & if\, c=0, \\ e\left(\frac{br^2}{2N}\right)\left(\frac{2Nc}{d}\right) & if\, c\neq 0. \end{cases} \tag{23}$$

We remark that in Proposition $1, \vartheta_{\frac{1}{2}+\nu}(N,r;\tau)$ are actually modular forms on a slightly larger congruence subgroup, but we have chosen to use $\Gamma_1(2N)$ for ease of exposition.

We next recall the structure of Fourier coefficients of positive index Jacobi forms for comparison. It is well known that holomorphic Jacobi forms have a theta decomposition involving the functions

$$\vartheta_{m,b}(z;\tau) := \sum_{\substack{\lambda\in\mathbb{Z} \\ \lambda\equiv b \pmod{2m}}} e^{\frac{\pi i\lambda^2\tau}{2m}+2\pi i\lambda z}.$$

The components of this decomposition are classical (vector-valued) modular forms [1]. The Fourier coefficients of meromorphic Jacobi forms of positive index are also understood. Specifically, in [2], Folsom and the first author, building on illuminating work of Dabholkar, Murthy, and Zagier [3] and Zwegers [6], considered the Kac-Wakimoto character of level (M,N) with $M>N$, $M,N\in 2\mathbb{N}$, which essentially corresponds to the meromorphic Jacobi form $\phi_{M,N}$ (the general case with $M>N$ is considered in [5]). These Kac-Wakimoto characters have a decomposition into a finite and a polar part, where the finite part has a theta decomposition similar to that of holomorphic Jacobi forms (but involving mock modular forms) and where the polar part is

$$\varphi^P(z;\tau) := -\sum_{j=1}^{\frac{N}{2}} \frac{D_{2j}(\tau)}{(2\pi i)^{2j-1}(2j-1)!}\frac{\partial^{2j-1}}{\partial u^{2j-1}}\left(e^{\pi i(M-N)u}\zeta^{-\frac{M-N}{2}}F_{M-N}\left(\frac{\tau}{2}-u,z-\frac{\tau}{2};\tau\right)\right)\Bigg|_{u=0}. \tag{24}$$

Here D_j is the j th Laurent coefficient of the level (M,N) Kac-Wakimoto character. Thus, we see that our functions $\phi_{M,N}$ have decompositions which are strikingly similar to the decompositions of positive index Jacobi forms, although in our case there are no associated 'finite parts'. As mentioned in Remark 1 following Theorem 3, this has an interesting interpretation in physics.

4.2 Quantum modular forms

In this section, we recall some definitions and examples of quantum modular forms and describe the quantum sets in Theorem 5. We begin with a few definitions (see [42] for additional background on quasimodular forms).

Definition.

A function f:H→C is an *almost holomorphic modular form of weight k* on a congruence subgroup *Γ* if it transforms as a modular form of weight *k* for *Γ* and is a polynomial in $\frac{1}{\mathrm{Im}(\tau)}$ with coefficients which are holomorphic on $\mathbb{H} \cup \mathbb{P}_1(\mathbb{Q})$. Moreover, *f* is a *quasimodular form of weight k* if it is the constant term of an almost holomorphic modular form of weight *k*.

Quantum modular are then defined as follows (see [21] for background on quantum modular forms).

Definition.

For any infinite 'quantum set' Q⊆Q, we say a function f:Q→C is a *quantum modular form* of weight *k* on a congruence subgroup *Γ* if for all *γ*∈*Γ*, the cocycle

$$G(a,b;\tau) := \sum_{n\geq 0}(-1)^n q^{\left(n+\frac{a}{b}\right)^2}.$$

extends to an open subset of and is analytically 'nice'. Here nice could mean continuous, smooth, real-analytic, etc. We say that *f* is a *strong quantum modular form* if there is a formal power series over attached to each point in with a stronger modularity requirement (see [21]).

Remark.

All of the quantum modular forms occurring in this paper have cocycles defined on which are real-analytic except at one point. Moreover, they have full asymptotic expansions towards rational points in their quantum sets which agree with the asymptotic expansions of mock modular forms defined on the lower half plane.

Especially relevant for us are certain partial theta functions which were shown to be quantum modular forms in recent work of Folsom, Ono, and Rhoades [23], namely,

$$G(a,b;\tau) := \sum_{n\geq 0}(-1)^n q^{\left(n+\frac{a}{b}\right)^2}.$$

For any a,b∈Z with $(a,b)=1$, $a>0$, define the following quantum set, where all fractions are assumed to have coprime denominator and numerator throughout

$$\mathcal{Q}_{a,b} := \left\{ \frac{h}{k} \in \mathbb{Q} : h > 0,\ b|2h,\ b \nmid h,\ k \equiv a \pmod{b},\ k \geq a \right\}.$$

Since for $r = \frac{j}{2} \in \frac{1}{2} + \mathbb{Z}$

$$\Theta_{\frac{1}{2}}(N, r; \tau) = G\left(j, 2N; \frac{N\tau}{2}\right),$$

it suffices to study the quantum modular properties of $G(a,b;\tau)$. Although $a=0$ is excluded, it is easy to handle this case directly. Note that G(0,1;τ2) is essentially a modular form as $G(0,1;\tau) = \frac{\eta(\tau)^2}{2\eta(2\tau)} + \frac{1}{2}$ and also that$G(0,2N;\tau)=G(0,1;2N\ \tau)$. It is clear that $G\left(0,1;\frac{\tau}{2}\right)$ is quantum modular at any cusps where the eta quotient vanishes, namely, for $\tau \in \left\{ \frac{h}{k} \in \mathbb{Q} : k \equiv 1 \pmod{2} \right\}$. For $a>0$, the situation is more subtle. Folsom, Ono, and Rhoades proved that $G(a,b;\tau)$ have the following quantum properties:

Theorem 6([23]).

For b even, $G(a,b;\tau)$ is a strong quantum modular form of weight 1/2 with quantum set Qa,b.

Remark.

Although [23] only states the theorem for $0<a<b$, an inspection of the proof shows that it is true for general integers $(a,b)=1$ with $a>0$ and b are even.

When N is even, we also have the analogous weight $\frac{3}{2}$ partial theta functions $\Theta_{\frac{3}{2}}(N, r; \tau)$ (see Theorem 5).

The Fourier coefficients and partial theta functions of A_N

In this section, we prove Theorem 2.

Proof of Theorem 2.

Define a subdomain of Y

$$X := \{h \in Y|\mathrm{Re}(C(h)) > \mathrm{Re}(\alpha(h)) > 0 \,\forall \alpha \in \Delta^{+}\}$$

so that in particular $|\mathfrak{q}(x)| < |e^{(\delta-\varepsilon_{N+1})(x)}| < 1$ for all x in X. We begin with the following crucial lemma.

Lemma 5.

As a function in X, we have

$$e^{\widehat{\rho}}\widehat{R} = e^{Nd-\rho_1} q^{-\frac{(N+1)(N+2)}{24}} \sum_{r \in \mathbb{Z}} \mathrm{sign}(-r)(-1)^r \mathfrak{q}^{-\frac{r^2}{2(N+1)}} \mathfrak{q}^{\frac{r}{2}} e^{r\frac{2\rho_1}{N+1}} \sum_{t \in T_r} \sum_{w \in W^{\sharp}} \sigma(w) w \left(e^t \mathfrak{q}^{\frac{t^2}{2N}}\right).$$

Proof.

Inserting the statement of Lemma 1 into the one of Lemma 2 gives

$$e^{\widehat{\rho}}\widehat{R} = \sum_{\alpha \in A_N} t_\alpha \left(e^{\widehat{\rho}}R\right) = \sum_{\alpha \in A_N} t_\alpha \left(\sum_{w \in W^{\sharp}} \sigma(w) w \left(\frac{e^{\widehat{\rho}}}{(1+e^{\delta-\varepsilon_{N+1}})}\right)\right).$$

Using (16) and the bijectivity of the map $w: A_N \to A_N$ for every $w \in W^{\#}$, we get

$$e^{\widehat{\rho}}\widehat{R} = \sum_{\alpha \in A_N} \sum_{w \in W^{\sharp}} \sigma(w) w \left(t_{w^{-1}(\alpha)} \left(\frac{e^{\widehat{\rho}}}{(1+e^{\delta-\varepsilon_{N+1}})}\right)\right) = \sum_{\alpha \in A_N} \sum_{w \in W^{\sharp}} \sigma(w) w \left(t_\alpha \left(\frac{e^{\widehat{\rho}}}{(1+e^{\delta-\varepsilon_{N+1}})}\right)\right).$$

Let $\alpha = m_1\alpha_1 + \cdots + m_N\alpha_N$ be an element of A_N, and set $m_0 := m_{N+1} := 0$. By (15), we have

$$t_\alpha(\widehat{\rho}) = \widehat{\rho} + N\alpha - \left(\sum_{n=1}^{N+1} \left(m_n + \frac{N}{2}(m_n - m_{n-1})^2\right)\right) C \quad \text{and}$$

$$t_\alpha(\delta - \varepsilon_{N+1}) = \delta - \varepsilon_{N+1} - m_N C.$$

Hence,

$$e^{\widehat{\rho}}\widehat{R} = \sum_{m \in \mathbb{Z}^N} \sum_{w \in W^{\sharp}} \sigma(w) w \left(e^{\widehat{\rho}} \prod_{n=1}^{N+1} e^{N m_n \alpha_n} \mathfrak{q}^{m_n + \frac{N}{2}(m_n - m_{n-1})^2} \left(1 + e^{\delta-\varepsilon_{N+1}} \mathfrak{q}^{m_N}\right)^{-1}\right),$$

where we used the short-hand notation $m = (m_1, \ldots, m_N)$ and kept as before $m_0 = m_{N+1} = 0$. Recall that $\alpha_n = \varepsilon_n - \varepsilon_{n+1}$. We split the exponential of the affine Weyl vector as

$$e^{\widehat{\rho}} = e^{Nd-\rho_1} \prod_{n=1}^{N+1} e^{\frac{(N+2-2n)}{2}\varepsilon_n}.$$

Note that $N\, d-\rho_1$ is invariant under $W^{\#}$. Letting $q_n := m_n - m_{n-1}$, we then find the identity

$$\sum_{n=1}^{N+1} \left(\frac{N}{2} q_n^2 + m_n\right) = \frac{1}{2N} \sum_{n=1}^{N+1} \left(N\left(q_n + \frac{1}{2}\right) + (1-n)\right)^2 - \frac{(N+1)(N+2)}{24}.$$

Defining the set

$$S := \left\{ (s_1, \ldots, s_{N+1}) \in \frac{1}{2}\mathbb{Z}^{N+1} \;\middle|\; s_n = N\left(q_n + \frac{1}{2}\right) + (1-n), q_n \in \mathbb{Z}, \sum_{n=1}^{N+1} q_n = 0 \right\},$$

we obtain

$$e^{\widehat{\rho}}\widehat{R} = e^{Nd-\rho_1} \mathfrak{q}^{-\frac{(N+1)(N+2)}{24}} \sum_{s \in S} \sum_{w \in W^{\sharp}} \sigma(w) w \left(\prod_{n=1}^{N+1} e^{s_n \varepsilon_n} \mathfrak{q}^{\frac{s_n^2}{2N}} \left(1 + e^{\delta - \varepsilon_{N+1}} \mathfrak{q}^{-\frac{s_{N+1}}{N} - \frac{1}{2}}\right)^{-1} \right).$$

In the domain X, we can expand in a geometric series to find that $e^{\widehat{\rho}}\widehat{R}$ equals

$$e^{Nd-\rho_1} \mathfrak{q}^{-\frac{(N+1)(N+2)}{24}} \left(\sum_{\substack{s \in S \\ s_{N+1} \le -\frac{1}{2}}} \sum_{r=0}^{\infty} (-1)^r \sum_{w \in W^{\sharp}} \sigma(w) w \left(e^{r(\delta - \varepsilon_{N+1})} \mathfrak{q}^{-r\left(\frac{s_{N+1}}{N} + \frac{1}{2}\right)} \prod_{n=1}^{N+1} e^{s_n \varepsilon_n} \mathfrak{q}^{\frac{s_n^2}{2N}} \right) \right.$$
$$\left. - \sum_{\substack{s \in S \\ s_{N+1} > -\frac{1}{2}}} \sum_{r=1}^{\infty} (-1)^r \sum_{w \in W^{\sharp}} \sigma(w) w \left(e^{-r(\delta - \varepsilon_{N+1})} \mathfrak{q}^{r\left(\frac{s_{N+1}}{N} + \frac{1}{2}\right)} \prod_{n=1}^{N+1} e^{s_n \varepsilon_n} \mathfrak{q}^{\frac{s_n^2}{2N}} \right) \right).$$

Since the double sum converges absolutely in the domain X, we can interchange summations. Define

$$g_r := \begin{cases} \displaystyle\sum_{\substack{s \in S \\ s_{N+1} \le -\frac{1}{2}}} \sum_{w \in W^{\sharp}} \sigma(w) w \left(e^{-r(\delta - \varepsilon_{N+1})} \mathfrak{q}^{r\left(\frac{s_{N+1}}{N} + \frac{1}{2}\right)} \prod_{n=1}^{N+1} e^{s_n \varepsilon_n} \mathfrak{q}^{\frac{s_n^2}{2N}} \right) & \text{if } r \le 0, \\ \displaystyle\sum_{\substack{s \in S \\ s_{N+1} > -\frac{1}{2}}} \sum_{w \in W^{\sharp}} \sigma(w) w \left(e^{-r(\delta - \varepsilon_{N+1})} \mathfrak{q}^{r\left(\frac{s_{N+1}}{N} + \frac{1}{2}\right)} \prod_{n=1}^{N+1} e^{s_n \varepsilon_n} \mathfrak{q}^{\frac{s_n^2}{2N}} \right) & \text{if } r > 0. \end{cases}$$

Then

$$e^{\widehat{\rho}}\widehat{R} = e^{Nd-\rho_1}\mathfrak{q}^{-\frac{(N+1)(N+2)}{24}}\sum_{r\in\mathbb{Z}}(-1)^r\operatorname{sign}(-r)g_r.$$

We can express $\varepsilon_{N+1}-\delta$ in terms of the odd Weyl vector and positive even simple roots:

$$\varepsilon_{N+1}-\delta = -\delta + \frac{1}{N+1}\left(\sum_{n=1}^{N+1}\varepsilon_n + \sum_{n=1}^{N} n(\varepsilon_{n+1}-\varepsilon_n)\right) = \frac{2}{N+1}\rho_1 + \varepsilon_{N+1} - \frac{1}{N+1}\sum_{n=1}^{N+1}\varepsilon_n. \tag{25}$$

We see that $\varepsilon_{N+1}-\delta-\frac{2}{N+1}\rho_1$ is in $\frac{1}{N+1}A_N$. For $(s_1,\ldots,s_{N+1})\in S$, we find that

$$e^{-r(\delta-\varepsilon_{N+1})}\mathfrak{q}^{r\left(\frac{s_{N+1}}{N}+\frac{1}{2}\right)}\prod_{n=1}^{N+1}e^{s_n\varepsilon_n}\mathfrak{q}^{\frac{s_n^2}{2N}} = e^{r\frac{2\rho_1}{N+1}}\mathfrak{q}^{-\frac{r^2}{2(N+1)}}\mathfrak{q}^{\frac{r}{2}}\prod_{n=1}^{N+1}e^{t_n\varepsilon_n}\mathfrak{q}^{\frac{t_n^2}{2N}} \tag{26}$$

Combining (25) and (26), we can rewrite

$$e^{-r(\delta-\varepsilon_{N+1})}\mathfrak{q}^{r\left(\frac{s_{N+1}}{N}+\frac{1}{2}\right)}\prod_{n=1}^{N+1}e^{s_n\varepsilon_n}\mathfrak{q}^{\frac{s_n^2}{2N}} = e^{r\frac{2\rho_1}{N+1}}\mathfrak{q}^{-\frac{r^2}{2(N+1)}}\mathfrak{q}^{\frac{r}{2}}\prod_{n=1}^{N+1}e^{t_n\varepsilon_n}\mathfrak{q}^{\frac{t_n^2}{2N}}$$

with $t_n := s_n - \frac{r}{N+1} + r\delta_{n,N+1}$. Then

$$\sum_{n=1}^{N+1}t_n\varepsilon_n = \sum_{n=1}^{N+1}\sum_{j=1}^{n}t_j(\varepsilon_j-\varepsilon_{j+1}) = \sum_{n=1}^{N}\sum_{j=1}^{n}t_j\alpha_n.$$

Here we used that $t_1+\cdots+t_{N+1}=0$, which follows from the same property for the s_n. Let q_j be as in the definition of the set S; in particular, we can write $q_j=m_j-m_{j-1}$ with integers m_j for $1\le j\le N$, and $m_{N+1}=0$. Then

$$\sum_{j=1}^{n} t_j = -\frac{rn}{N+1} + \sum_{j=1}^{n} s_j = \frac{n(N-n+1)}{2} - \frac{rn}{N+1} + \sum_{j=1}^{n} Nq_j$$
$$= \frac{(N-n+1)n}{2} - \frac{rn}{N+1} + Nm_n.$$

Using the sets $T_r(5)$, we finally get

$$g_r = e^{r\frac{2\rho_1}{N+1}} \mathfrak{q}^{-\frac{r^2}{2(N+1)}} \mathfrak{q}^{\frac{r}{2}} \sum_{t \in T_r} \sum_{w \in W^\ddagger} \sigma(w) w \left(e^t \mathfrak{q}^{\frac{t^2}{2N}} \right).$$

Letting $\mathfrak{z} = -e^{\frac{2\rho_1}{N+1}}$, we deduce the following.

Corollary 1.

The identity A=B C holds as functions on X, where

$$A := \prod_{j=1}^{N+1} \frac{\mathfrak{q}^{\frac{1}{24}}}{(-e^{\delta-\varepsilon_j};\mathfrak{q})_\infty(-e^{\varepsilon_j-\delta}\mathfrak{q};\mathfrak{q})_\infty}, \qquad B := \frac{\mathfrak{q}^{-\frac{(N+1)^2}{24}}}{(\mathfrak{q};\mathfrak{q})_\infty^{N+1} \prod_{\alpha \in \Delta_0} (e^\alpha \mathfrak{q};\mathfrak{q})_\infty (e^{-\alpha}\mathfrak{q};\mathfrak{q})_\infty},$$

$$C := \sum_{r \in \mathbb{Z}} \operatorname{sign}(-r) \mathfrak{z}^r \mathfrak{q}^{-\frac{r^2}{2(N+1)}} \mathfrak{q}^{\frac{r}{2}} \sum_{t \in T_r} v_t \mathfrak{q}^{\frac{t^2}{2N}}.$$

Proof.

The corollary follows immediately from Lemma 5 by inserting v_t in the definition of C in (18).

Evaluating the expressions in this equality provides a nice expansion of $\phi_{N+1}(z;\tau)$.

Corollary 2.

Inside the range $|q|<|\zeta|<1$, *we have*

$$\phi_N(z;\tau) = \frac{i^N}{d_{N-1}\eta(\tau)^{N^2+N}} \sum_{r \in \mathbb{Z}} \zeta^{r+\frac{N}{2}} \operatorname{sign}(r) q^{-\frac{\left(r+\frac{N}{2}\right)^2}{2N}} \sum_{t \in T_r} \prod_{\alpha \in \Delta_0^+} (t|\alpha) q^{\frac{t^2}{2(N-1)}}.$$

Proof.

The evaluation v maps every regular exponential e^λ for $\lambda \in \frac{1}{N} A_{N-1}$ to 1. The application of v to A and B is finite for $|\mathfrak{q}(x)| < |\mathfrak{z}^{-1}(x)| < 1$ and the same is true for C by Lemma 3. The identity (25) implies that $\nu\left(e^{\delta-\varepsilon_j}\right) = e^{-\frac{2\rho_1}{N}} = -\zeta$ for all

$j = 1, \ldots, N$, so that

$$\nu(A) = \mathfrak{q}^{\frac{N}{24}}\left(\mathfrak{z}^{-1};\mathfrak{q}\right)_\infty^{-N}(\mathfrak{z}\mathfrak{q};\mathfrak{q})_\infty^{-N} \quad \text{and} \quad \nu(B) = \mathfrak{q}^{-\frac{N^2}{24}}(\mathfrak{q};\mathfrak{q})_\infty^{-N^2}.$$

By Lemma 3,

$$\nu(\nu_t) = \frac{1}{d_{N-1}} \prod_{\alpha\in\Delta_0^+} (t|\alpha)$$

and the evaluation $\nu(C)$ follows. All three evaluations $\nu(A)$,$\nu(B)$,$\nu(C)$ are meromorphic functions on $\left\{x = -2\pi i\tau d + \frac{4\pi i z h_{\rho_1}}{N-1}\right.:$ so that the result follows with $\zeta = \mathfrak{z}^{-1}(x)$ and $q = \mathfrak{q}(x)$.

This completes the proof as Corollary 1 and Lemma 3 imply Theorem 2. □

The case N=1 can be proven in a very similar manner using (9), which is the denominator identity of $\widehat{g\ell}(1|1)$ (see Example 4.1 of [16]).

Example 1.

The Fourier coefficients of $\phi_1(z;\tau)$ are given by

$$\chi(1,r;\tau) = \frac{iq^{-\frac{r^2}{2}}}{\eta(\tau)^3}\sum_{m=0}^{\infty}(-1)^m q^{\frac{\left(m+\left|r-\frac{1}{2}\right|+\frac{1}{2}\right)^2}{2}}.$$

Proof.

Suppose $|q|<|\zeta|<1$. Expanding (9) in a geometric series and rewriting easily gives the statement.

4

INDECOMPOSABLE DECOMPOSITION AND COUNISERIAL DIMENSION

A. Ghorbani[1], S. K. Jain[2], Z. Nazemian[1]

[1] Department of Mathematical Sciences, Isfahan University of Technology, Isfahan, Iran

[2] Department of Mathematics, Ohio University, Athens, OH 45701, USA

KEYWORDS

niform module; Maximal right quotient ring; Indecomposable decomposition; Uniserial dimension; Couniserial dimension; Von Neumann regular ring; Semisimple module

1 INTRODUCTION

In this article we introduce a notion of dimension of a module, to be called couniserial dimension. It is an ordinal valued invariant that is in some sense a measure of how far a module is from being uniform. In order to define couniserial dimension for modules over a ring R, we first define, by transfinite induction, classes ζα of R -modules for all ordinals α≥1. First we remark that if a module MM is isomorphic to all its non-zero submodules, then M must be uniform. To start with, let ζ1 be the class of all uniform modules. Next, consider an ordinal α>1; if ζβ has been defined for all ordinals β<α, let ζα be the class of those RR-modules M such that for every non-zero submodule N of M, where N≇M, we have N∈⋃β<αζβ. If an R-module M belongs to some ζα, then the least such αα is called the *couniserial dimension* of MM, denoted by c.u.dim (M). For M=0, we define c.u.dim(M)=0. If a non-zero module M does not belong to any ζα, then we say that c.u.dim (M) is not defined, or that M has no couniserial dimension. Equivalently, Proposition 3.3 shows that an R-module M has couniserial dimension if and only if for each descending chain of submodules of M, $M_1 \geq M_2 \geq \cdots$, there exists n≥1, either Mn is uniform or Mn≅Mk for all k≥n. It is clear by the definition that every submodule and so every summand of a module with couniserial dimension has couniserial dimension. Also note that, for the integer number nn, couniserial dimension of Z^n is n. An example is given to show that the direct sum of two modules each with couniserial dimension (even copies of a module) need not have couniserial dimension. In Sect. 3, we prove some basic properties of the couniserial dimension. In Sect. 4, we prove our main results. It is shown in Theorem 4.3 that a module of countable (finite or infinite) couniserial dimension can be decomposed into indecomposable modules. Theorem 4.5shows that a Dedekind finite module with couniserial dimension is a finite direct sum of indecomposable modules. Theorem 4.10 in Sect. 4 shows that for a semiprime right non-singular ring R with maximal right quotient ring Q, if QR has couniserial dimension, then R is a semiprime right Goldie ring which is a finite product of piecewise domains. The reader may compare this with the well known result that a prime ring with Krull dimension is a right Goldie ring but need not be a piecewise domain. Furthermore, a prime right Goldie ring need not have couniserial dimension as is also the case for Krull dimension.

In Sect. 5, we give some applications of couniserial dimension. It is shown in Proposition 5.2 that a module MM with finite length is semisimple if and only if for every submodule N of M the right R-module $\oplus_{i=1}^{\infty} M/N$ has couniserial dimension. As a consequence a commutative noetherian ring R is semisimple if and only if for every finite length module MM the module $\oplus_{i=1}^{\infty} M$ has couniserial dimension. It is shown in Proposition 5.4 that if PP is an anti-coHopfian projective right R-module and $\oplus_{i=1}^{\infty} E(P)$ has couniserial dimension, then PP is injective. As another application we show that all right (left) R-module have couniserial dimension if and only if RR is semisimple artinian (see Theorem 5.8). Several examples are included in the paper that demonstrates as to why the conditions imposed are necessary.

2 DEFINITIONS AND NOTATIONS

Recall that a semisimple module M is said to be *homogeneous* if M is a direct sum of pairwise isomorphic simple submodules. A module M has *finite uniform dimension* (or*finite Goldie rank*) if M contains no infinite direct sum of non-zero submodules, or equivalently, there exist independent uniform submodules U1,...,Un in M such that $\oplus_{i=1}^{n} U_i$ is an essential submodule of M. Note that n is uniquely determined by M. In this case, it is written u.dim(M)=n.

For any module M, we define $Z(M) = \{x \in M$: an (x) is an essential right ideal of R}. It can be easily checked that Z(M) is a submodule of M. If Z(M)=0, then M is called a*non-singular* module. In particular, if we take M=R, then R is called right non-singular if $Z(R_R)=0$.

A ring R is called a *right Goldie* ring if it satisfies the following two conditions: (i) R has ascending chain condition on right annihilator ideals and, (ii) u.dim(R_R) is finite. Recall that a ring R is *right V-ring* if all right simple R-modules are injective. A ring R is called *fully right idempotent* $I = I^2$, for every right ideal II. We recall that a right V-ring is fully right idempotent (see [19, Corollary 2.2]) and a prime fully right idempotent ring is right non-singular (see [2, Lemma 4.3]). So a prime right V-ring is right non-singular. Recall that a module M is called Σ-*injective* if every direct sum of copies of M is injective. A ring R is called *right* Σ-*V-ring* if each simple right module is Σ-injective.

In this paper, for a ring R, $Q=Q_{max}(R)$ stands by the maximal right quotient ring RR. It is well known that if RR is a right non-singular, then the injective hull of R_R, E(R_R), is a ring and is equal to the maximal right quotient ring of RR [9, Corollary 2.31].

A module M is called *Hopfian* if M is not isomorphic to any of its proper factor modules (equivalently, every onto endomorphism of M is 1-1). *Anti-Hopfian* modules are introduced by Hirano and Mogami [13]. Such modules are isomorphic to all their non-zero factor modules. A module M is called uniserial if the lattice of submodules are linearly ordered. Anti-Hopfian modules are uniserial artinian.

Recall that a module M is called *coHopfian* if it is not isomorphic to a proper submodule (equivalently, every 1-1 endomorphism of M is onto). Varadarjan [22] dualized the concept of anti-Hopfian module and called it anti-coHopfian module. With slight modification we will call a non-zero module to be *anti-coHopfian* if it is isomorphic to all its non-zero submodules. A non-zero module M is called *uniform* if the intersection of any two non-zero submodules is non-zero. We see an anti-coHopfian module is noetherian and uniform (see Lemma 3.1 below).

An R-module M has cancellation property if for every R-modules N and T, $M\oplus N\cong M\oplus T$ implies $N\cong T$. Every module with semilocal endomorphism ring has cancellation property [16]. Since the endomorphism ring of a simple module is a division ring, it has cancellation property.

Unless otherwise specified, throughout this paper R denotes an arbitrary ring and all modules are assumed to be unitary right modules. If N is a submodule (resp. proper submodule) of M we write $N\leq M$ (resp. $N<M$). Also, for a module M, $\oplus_{i=1}^{\infty}M$ stands for countably infinite direct sum of copies of M. If N is a submodule of M and $k>1$, $\oplus_{i=k}^{\infty}N=\oplus_{i=1}^{\infty}N_i$ is a submodule of $\oplus_{i=1}^{\infty}M$ with $N_1=N_2=\cdots=N_{k-1}=0$ and for $i\geq k$ $N_i=N$.

3 BASIC AND PRELIMINARY RESULTS

As defined in the introduction, couniserial dimension is an ordinal valued number. The reader may refer to [21] regarding ordinal numbers. We begin this section with a lemma and a remark on the definition of couniserial dimension.

Lemma 3.1

An anti-coHopfian module is uniform noetherian.

Proof

Since M is isomorphic to each cyclic submodules, M is cyclic and every submodule of M is cyclic and so M is noetherian. Thus M has a uniform submodule, say U. Since $U\cong M$, M is uniform.

Remark 3.2

We make the convention that a statement "c.u.dim $(M)=\alpha$" will mean that the couniserial dimension of M exists and equals $\alpha\alpha$. By the definition of couniserial dimension, if M has couniserial dimension and N is a submodule of M, then N has couniserial dimension and c.u.dim(N)≤ c.u.dim(M). Moreover, if M is not uniform and c.u.dim(M)= c.u.dim(N), where N is a submodule of M, then $M\cong N$. On the other hand, since every set of ordinal numbers has supremum, it follows immediately from the definition that M has couniserial dimension if and only if for all submodules N of M with N≇M, c.u.dim(N) is defined. In the latter case, if $\alpha = \sup\{\text{c.u.dim}(N) \mid N \leq M, N \ncong M\}$, then $\text{c.u.dim}(M) \leq \alpha + 1$.

The next proposition provides a working criterion for a module to have couniserial dimension.

Proposition 3.3

An RR-module MM has couniserial dimension if and only if for every descending chain of submodules $M_1 \geq M_2 \geq \cdots$, there exists n≥1 such that MnMn is uniform or $M_n \cong M_k$ for all $k \geq n$.

Proof

(⇒) Let $M_1 \geq M_2 \geq \cdots$ be a descending chain of submodules of M. Put γ= inf {c.u.dim(M_n) | n≥1. So γ= c.u.dim (Mn) for some n≥1. If Mn is not uniform, then Mn≅Mk for all k≥n, because γ is infimum.

(⇐) If MM does not have couniserial dimension, then M is not uniform and so there exists a submodule M1 of M such that M1≇M and M1 does not have couniserial dimension, by the above remark. So there exists a submodule M2 of M1 such that M2≇M1 and M2does not have couniserial dimension. Continuing in this manner, we obtain a descending chain of submodules M1≥M2≥⋯, such that for every i≥1, Mi does not have couniserial dimension and Mi≇Mi+1, a contradiction. This completes the proof. □

As a consequence, we have the following corollary.

Corollary 3.4

Every artinian module has couniserial dimension.

Lemma 3.5

If M is an R-module and c.u.dim(M)=α, then for any 0≤β≤α, there exists a submodule N of M such that c.u.dim(N)=β.

Proof

The proof is by transfinite induction on c.u.dim(M)=α. The case α=1 is clear. Let α>1 and 0≤β<α, then, using Remark 3.2, there exists a submodule K of M such that K≇M and β≤ c.u.dim(K). Now since β≤ c.u.dim(K)<α, by induction hypothesis, there exists a submodule N of K such that c.u.dim(N)=β. □

As a consequence we have the following.

Lemma 3.6

Every module with couniserial dimension has a uniform submodule.

In the next proposition we observe that every module of finite couniserial dimension has finite uniform dimension.

Lemma 3.7

Let MM be an RR-module of finite couniserial dimension. Then M has finite uniform dimension and u.dim(M)≤ c.u.dim(M).

Proof

The proof is by induction on c.u.dim(M)=n. The case n=1 is clear. Let n>1 and N be a submodule of M such that c.u.dim(N)=n−1. Thus by the inductive hypothesis, N has finite uniform dimension. Put m=m= u.dim(N). If N is not essential in M, then there exists a uniform submodule U of M such that $N \cap U = 0$. Thus $N \oplus U$ is a submodule of M of uniform dimension m+1. Then $(N \oplus U) \ncong N$ and so n−1< c.u.dim $(N \oplus U) \leq n$. Thus $(N \oplus U) \cong M.$, by Remark 3.2. This proves the lemma. □

Example 3.8

There exist modules of infinite couniserial dimension but of finite uniform dimension. Take $M = \mathbb{Z}_{p^\infty} \oplus \mathbb{Z}_{p^\infty}$. Then M is an artinian Z-module of infinite couniserial dimension but of finite uniform dimension 2.

In the following, we show that, in a special case, equality holds in the previous lemma.

Lemma 3.9

Let M be an injective non-uniform R-module of finite couniserial dimension. Then c.u.dim(M)= u.dim(M) if and only if M is a finitely generated semisimple module.

Proof

(⇐) is clear. (⇒). Let c.u.dim(M)= u.dim(M)=m>1. Then $M = E_1 \oplus \cdots \oplus E_m$, where Ei are uniform injective modules. If E_1 is not simple then there exists a non-injective submodule K of E_1. Thus $K \oplus E_2 \oplus \cdots \oplus E_m$ is not isomorphic to M. But clearly c.u.dim$(K \oplus E_2 \oplus \cdots \oplus E_m) \geq m$. a contradiction. This completes the proof. □

Note that the condition of being injective is necessary in the above proposition.

Example 3.10

We can see easily that for $M = \mathbb{Z} \oplus \mathbb{Z}$., c.u.dim(M)= u.dim(M) =2 but M is not semisimple. Also, the next lemma shows that there exists a module of finite uniform dimension without couniserial dimension.

The following lemma shows that the direct sum of two uniform modules may not have couniserial dimension.

Lemma 3.11

Let D be a domain and SS be a simple D-module. If S⊕D as D-module has couniserial dimension, then D is a principal right ideal domain.

Proof

Let I be a non-cyclic right ideal of D. Choose a non-zero element x∈I. Set J1=xR which is isomorphic to D. Thus there exists a right ideal J2 of D such that J2≅I and J2≤J1. Now let J3 be a cyclic right ideal contained in J2 and by continuing this manner we have a descending chain J1≥J2≥⋯ of right ideals of D where for each odd integer ii, Ji is cyclic and for each even integer i, Ji is not cyclic. Now consider the descending chain S⊕J1≥⋯S⊕J2≥⋯ of submodules of S⊕D. Since SS has cancellation property and for each ii, S⊕Ji is not uniform, by using Proposition 3.3, we see that, for some nn, Jn≅Jn+1, a contradiction. Thus D is a principal right ideal domain. □

Remark 3.12

1) The simple module SS in the statement of the Lemma 3.11 can be replaced by any cancellable module. Indeed it follows from Theorem 4.10, proved later that, if the maximal right quotient ring Q of a domain D has couniserial dimension as D-module, then QD has cancellation property and so if Q⊕D as D-module has couniserial dimension, D must be right principal ideal domain.
2) Also, since a Dedekind domain has cancellation property, similar proof shows that if D is a Dedekind domain which is not right principal ideal domain, then D⊕D does not have couniserial dimension. This example shows that even direct sum of a uniform module with itself may not have couniserial dimension.

The definition of addition two ordinal numbers can be given inductively. If αα and β are two ordinal numbers then $\alpha + 0 = \alpha, \alpha + (\beta + 1) = (\alpha + \beta) + 1$ and if γ is a limit ordinal then α+γ is the limit of α+β for all β<γ (see [21]).

Lemma 3.13

(See [21, Theorem 7.10]) For ordinal numbers αα, ββ and γγ, we have the following:

If $\alpha < \beta$, *then* $\gamma + \alpha < \gamma + \beta$.
If $\alpha < \beta$, *then* $\alpha + \gamma \leq \beta + \gamma$.

We call an R-module M*f ully coHopfian* if every submodule of M is coHopfian. Note that artinian modules are fully coHopfian. If I is the set of prime numbers, then ⊕p∈IZp is an example of fully coHopfian Z-module that it is not artinian.

Proposition 3.14

Let $M = M_1 \oplus M_2$ be a fully coHopfian R-module with couniserial dimension. Then c.u.dim(M)≥ c.u.dim (M1)+ c.u.dim(M2).

Proof

We may assume M1,M2≠0. We use transfinite induction on c.u.dim(M2)=α. Since M1≇M, c.u.dim(M)≥ c.u.dim(M1)+1. So the case α=1 is clear. Thus, suppose α>1 and for every right R-module L of couniserial dimension less than αα, c.u.dim(M1⊕L)≥ c.u.dim(M1)+ c.u.dim(L). There are two possibilities: α is a successor ordinal, or a limit ordinal. Suppose first that αα is a successor ordinal, then there exists an ordinal number γ such that c.u.dim(M2)=γ+1. Using Lemma 3.5, there exists a non-zero submodule K of M2 such that c.u.dim(K)=γ<α. So by the induction hypothesis

$$\text{c.u.dim}(M_1) + \gamma = \text{c.u.dim}(M_1) + \text{c.u.dim}(K) \leq \text{c.u.dim}(M_1 \oplus K).$$

Using our assumption and Remark 3.2, we have c.u.dim(M1⊕K)< c.u.dim(M) and hence c.u.dim(M1)+ c.u.dim(M2).

If α is a limit ordinal and 1≤β<α, then by Remark 3.2, there exists a non-zero submodule K of M2 such that β≤ c.u.dim(K). Then by induction hypothesis $\text{c.u.dim}(M_1)+\beta \leq \text{c.u.dim}(M_1)+\text{c.u.dim}(K) \leq \text{c.u.dim}(M_1 \oplus K) < \text{c.u.dim}(M)$. Therefore $\text{c.u.dim}(M_1) + \alpha = \sup\{\text{c.u.dim}(M_1) + \beta \mid \beta < \alpha\} \leq \text{c.u.dim}(M)$. □

The condition fully coHopfian of Proposition 3.14 is necessary.

Example 3.15

For the $\mathbb{Z}$-modules $M = \oplus_{i=1}^{\infty}\mathbb{Z}_p$, and $L = \mathbb{Z}_p$, we have $M \cong M \oplus L$. One can see $\text{c.u.dim}(M) = \omega$ and so $\text{c.u.dim}(M) \ngeq \text{c.u.dim}(M) + \text{c.u.dim}(L)$. Also, in general, we don't have the equality in Proposition 3.14. Consider the Z-module $M = \mathbb{Z}_2 \oplus \mathbb{Z}_4$. Then, M is fully coHopfian and 3= c.u.dim(M)> c.u.dim(Z2)+ c.u.dim(Z4).

Here we prove another result on fully coHopfian module:

Proposition 3.16

Let M be an RR-module and N be a cancellable module (for example a simple module) such that N⊕M has couniserial dimension. If M is fully coHopfian, then M is artinian.

Proof

Let M be fully coHopfian and let M1≥M2≥··· be a descending chain of submodules of M. Then $N \oplus M_1 \geq N \oplus M_2 \geq \cdots$ is a descending chain of submodules of N⊕M and so for some n, $M_i \cong M_n$ for each $i \geq n$. for each i≥n. Now since M is fully coHopfian, we have Mi=Mn for each i≥n. □

Let us recall the definition of uniserial dimension [20].

Definition 3.17

In order to define uniserial dimension for modules over a ring RR, we first define, by transfinite induction, classes ζα of R-modules for all ordinals α≥1. To start with, let ζ1 be the class of non-zero uniserial modules. Next, consider an ordinal α>1; if ζβ has been defined for all ordinals β<α, let ζα be the class of those RR-modules M such that, for every submodule N<M, where M/N≇M, we have *M/N*▯ ∪β<αζβ. If an RR-module M belongs to some ζα, then the least such αα is the *uniserial dimension of* M, denoted u.s.dim(M). For M=0, we define u.s.dim(M)=0. If M is non-zero and M does not belong to any ζα, then we say that "u.s.dim(M) is not defined," or that "M has no uniserial dimension."

Remark 3.18

Note that, in general, there is no relation between the existence of the uniserial dimension and the existence of the couniserial dimension of a module. For example, the polynomial ring in infinite number of commutative indeterminates over a field k, R=k[x1,x2,...] has this property that c.u.dim(R_R)=1 but R_R does not have uniserial dimension (see [6, Remark 2.3]). It follows by the definition that a semisimple module M has uniserial dimension if and only if M has couniserial dimension, in which case u.s.dim(M)=α if and only if c.u.dim(M)=α. Furthermore a semisimple module M has couniserial dimension if and only if M is a finite direct sum of homogeneous semisimple modules (see [20, Proposition 1.18]).

Using the above remark we have the following interesting results.

Corollary 3.19

All right semisimple modules over a ring RR have couniserial dimension if and only if there exist only finitely many non-isomorphic simple right RR-modules.

Lemma 3.20

Let MM be a finitely generated cancellable module. Then:

1) *If* $\oplus_{i=1}^{\infty} E(M)$ has couniserial dimension, then M is injective.
2) Suppose that M is simple. Then $\oplus_{i=1}^{\infty} E(M)$ has couniserial dimension if and only if M is injective.

Proof (1) Consider the descending chain

$$M \oplus (\oplus_{i=2}^{\infty} E(M)) \geq M^{(2)} \oplus (\oplus_{i=3}^{\infty} E(M)) \geq \cdots$$

of submodules of $\oplus_{i=1}^{\infty} E(M)$ where $M^{(n)} = \oplus_{i=1}^{\infty} M_i$ with $M_1 = \cdots = M_n = M$ and for each $i > n$, $M_i = 0$. Then, by Proposition 3.3, there exists n≥1 such that

$$M^{(n)} \oplus (\oplus_{i=n+1}^{\infty} E(M)) \cong M^{(n+1)} \oplus (\oplus_{i=n+2}^{\infty} E(M))$$

and so so $\oplus_{i=n+1}^{\infty} E(M) \cong M \oplus (\oplus_{i=n+2}^{\infty} E(M))$, because M is cancellable . Since M is finitely generated, there exists a right module L such that for some k, $E(M)^k \cong M \oplus L$. This shows M is injective.

(2) It is clear by part 1 and the statement in Remark 3.18.

4 MAIN RESULTS

In this section we use our basic results to prove the main results.

Proposition 4.1

Let M_R be an injective module and NRNR be a cancellable module over a commmutative ring R (for example a simple module) such that N⊕M has couniserial dimension. Then *M* is Σ-injective.

Proof

According to [15, Theorem 6.17], it is enough to show that RR satisfies the ascending chain condition on ideals of RR that are annihilators of subsets of M. Let $I_1 \leq I_2 \leq \cdots$ be a chain of such annihilator ideals. Then for each i, $M_i = \mathrm{ann}_M(I_i)$ is a submodule of M and so we have descending chain $N \oplus M_1 \geq N \oplus M_2 \geq \cdots$ of submodules of *N⊡M*. Then there exists a positive integer nn such that *MnMi* for all *i≥n*. Thus ann (Mi) = ann (Mn). Therefore for each $i \geq n$, $I_i = I_n$. □

Remark 4.2

If SS is a simple module, then S⊕S has couniserial dimension and S is also a cancellable module. Thus, the above result provides another proof for the fact that commutative V-rings (i.e, von Neumann regular rings) are Σ-V-ring. Now if R is a right V-ring that is not right Σ-V-ring, see [15, Example, page 60], then there exists a simple right R-module S which is not Σ-injective. This shows that the commutativity condition cannot be dropped in the above proposition.

The next result shows that if a module has countable couniserial dimension then it can be decomposed into indecomposable modules.

Theorem 4.3

For an R-module M, if c.u.dim(M)≤ω, then M has indecomposable decomposition.

Proof

The proof is by induction on c.u.dim(M)=α. The case α=1 is clear. If α>1 and M is not indecomposable, then $M = N_1 \oplus N_2$, where N1 and N2 are non-zero submodules of M. If c.u.dim(Ni)< c.u.dim(M), i=1,2, then by induction hypothesis M has indecomposable decomposition. If not, for definiteness let c.u.dim(N1)= c.u.dim(M). Then M≅N1, by Remark 3.2.

Thus it contains an infinite direct sum of uniform modules, say $\oplus_{i=1}^{\infty} K_i$. Clearly, c.u.dim$(\oplus_{i=1}^{\infty} K_i) \geq \omega$. Thus we have $M \cong \oplus_{i=1}^{\infty} K_i$. □

Remark 4.4

We do not know whether the above proposition holds for a module of arbitrary couniserial dimension. For infinite countable couniserial dimension one can show under some condition that the module can be represented as a direct sum of uniform modules.

Recall that a module *M* is called *Dedekind finite* if *M* is not isomorphic to any proper direct summand of itself. Clearly, every direct summand of a Dedekind finite module is a Dedekind finite module. Obviously, a Hopfian module is Dedekind finite. Since all finitely generated modules over a commutative ring are Hopfian (see [10]), they provide examples of Dedekind finite modules.

Theorem 4.5

If MM is a Dedekind finite module with couniserial dimension, then MM has finite indecomposable decomposition.

Proof

The proof is by induction on c.u.dim(M)=α. The case α=1 is clear. Let α>1 and every Dedekind finite module with c.u.dim less than αα be decomposed to finitely many indecomposable modules. If M is not indecomposable, then M=M1⊕M2M=. Since Mi≇M, using Remark 3.2, c.u.dim(Mi)< c.u.dim(M) and so, by induction hypothesis, Mi have finite indecomposable decomposition. This completes the proof. □

A ring R is called a *von Neumann regular ring* if for each x∈R, there exists y∈R such that xyx = x, equivalently, every principal right ideal is a direct summand. RR is *unit regular ring* if for each x∈R, there exists a unit element u∈R such that x=xux. As a consequence of the above theorem we have the following corollary.

Corollary 4.6

Every Dedekind finite von Neumann regular ring (in particular, unit regular rings) with couniserial dimension is semisimple artinian.

A ring RR is called a PWD (*piecewise domain*) if it possesses a complete set $\{e_i | 0 \leq i \leq n\}$ of or thogonal idempotents such that xy=0 implies x=0 or y=0 whenever x∈eiRek and y∈ekRej. Note that the definition is left-right symmetric and all eiRei are domain, see [12].

An element x of R is called regular if its right and left annihilators are zero.

Proposition 4.7

Let R be a semiprime right Goldie ring with couniserial dimension. If u.dim(RR)=n, then R has a decomposition into n uniform modules. In particular, it is a piecewise domain.

Proof

We can assume that $n > 1$. Let $I_1 = U_1 \oplus \cdots \oplus U_n$ be an essential right ideal of R. Then, by [8, Proposition 6.13], I1 contains a regular element x and thus J1=xR is a right ideal of R which is RR-isomorphic to R. So u.dim(J1)=n and it contains an essential right ideal I2 of R such that it is a direct sum of n uniform right ideals. By continuing in this manner we obtain a descending chain I1≥J1≥I2≥··· of right ideals of R such that Ii are direct sum of n uniform right ideals and JiJi are isomorphic to R. Since R has couniserial dimension, for some n, In≅R. The last statement follows from [12, Pages 2–3]. This completes the proof. □

Remark 4.8

There exists an example of simple noetherian ring of uniform dimension 22 which has no non-trivial idempotents (c.f. [18, Example 7.16, page 441]). So by the above proposition this provides an example of prime right Goldie ring without couniserial dimension.

Lemma 4.9

Let R be a right non-singular ring with maximal right quotient ring Q. Let M be a Q-module. If M is non-singular R-module, such that MR has couniserial dimension, then MQ has couniserial dimension.

Proof

Let M≥M1≥M2≥··· be a descending chain of Q-submodules of M. So it is a descending chain of R-submodules of M and thus, for some nn, Mn is uniform R-module or Mn≅Mi as RR-modules for all i≥n. If Mn is uniform R-module, then it is also uniform Q-module. So let Mn≅Mi as R-modules and let φi be this isomorphism. If q∈Q and t∈Mn there exists an essential right ideal E of R such that qE≤R. So $\varphi_i(tqE) = \varphi_i(tq)E$ and also $\varphi_i(tqE) = \varphi_i(t)qE$. Then $\varphi_i(tq)E = \varphi_i(t)qE$. Since QQ is right non-singular, $\varphi_i(tq) = \varphi_i(t)q$. Thus φ_i is a Q-isomorphism. This completes the proof.

A ring R is semiprime (prime) right Goldie ring if and only if its classical right quotient ring is semisimple (simple) artinian ring [9, Theorems 3.35 and 3.36]. Semiprime right Goldie rings are non-singular. A semiprime right non-singular ring RR is semiprime right Goldie ring if and only if u.dim (R_R) is finite, [9, Theorem 3.17]. Recall that a *right full linear ring* is the ring of all linear transformations (written on the left) of a right vector space over a division ring. If the dimension of the vector space is finite, a right full linear ring is exactly a simple artinian ring.

Theorem 4.10

Let R be a semiprime right non-singular ring with maximal right quotient ring, Q. If Q as an R-module has couniserial dimension, then R is a semiprime right Goldie ring which is a finite product of prime Goldie rings, each of which is a piecewise domain.

Proof

It is enough to show that R has finite uniform dimension. Since QR has couniserial dimension, R_R has couniserial dimension and so every right ideal of RR has couniserial dimension. Thus Lemma 3.6 implies that every right ideal contains a uniform submodule. Now by [9, Theorem 3.29] the maximal right quotient ring of RR is a product of right full linear rings, say $Q = \prod_{i\in I} Q_i$, where Q_i are right full linear rings. Note that since RRRR is right non-singular, QR is also non-singular and so, using Lemma 4.9, Q_Q has couniserial dimension. At first we claim each Qi is endomorphism ring of a finite dimensional vector space. Assume the contrary. Then Qj is the endomorphism ring of an infinite dimensional vector space, for some jj. Thus Qj≅Qj×Qj and so if ι:Qj⟶Q be the canonical embedding, then ι(Qj) is a right ideal of Q and there exists a Q-isomorphism $Q \cong \iota(Q_j) \times Q$. Then there exist right ideals T1 and T of Q such that Q=T1⊕T, T1 and Q are isomorphic as Q-modules and T≅ι(Qj) as Q-module. Because Qj is the endomorphism ring of an infinite dimensional vector space, it has a right ideal which is not principal, for example its socle. So ι(Qj) and thus TT contains a non-cyclic right ideal of Q and thus since T≅Q/T1, there exists a non-cyclic right ideal of Q, say K1 such thatQ≥K1≥T1. Now T1 is isomorphic to Q. So we can have a descending chain Q>K1>T1>K2>T2>⋯ of right ideals of Q such that Ti are cyclic but Ki are not cyclic. This is a contradiction. So all Qi are endomorphism ring of finite dimensional vector spaces. Now to show R is semiprime right Goldie ring it is enough to show that the index set II is finite. If II is infinite, there exist infinite subsets I1 and I2 of I such that I=I1∪I2. and I1∩I2 is empty. Let T1=∏i∈INi such that Ni=Qi for all i∈I1 and Ni=0 for all i∈I2. Similarly let T=∏i∈IMi such that Mi=Qi for all i∈I2 and Mi=0 for all i∈I1. Then T1 and T are right ideals of Q and Q=T1⊕T. T contains a right ideal of Q which is not cyclic, for example ⊕i∈IMi. Since T≅Q/T1, there exists a non-cyclic right ideal K1 of Q such that Q≥K1≥T1. Note that T1 is a cyclic Q-module and because I1 is infinite, the structure of T1 is similar to that of Q. We can continue in this manner and find a descending chain of right ideals of Q such that Ki are non cyclic Q-modules and Ti are cyclic Q modules, which is a contradiction. Therefore I is finite and R must have finite uniform dimension. This shows RR is semiprime right Goldie ring and so Proposition 4.7 and [12, Corollary 3] imply that it is a direct sum of prime right Goldie rings. □

We may point out that unlike a semiprime ring with right Krull dimension, a semiprime ring with couniserial dimension need not be a right Goldie ring. See Dubrovin [5] that contains an example of a primitive uniserial ring with non-zero nilpotent elements.

Next we show that the converse of the above theorem is not true, in general. In fact we show that there exists a prime right Goldie ring R such that c.u.dim(R_R)=2 and QR does not have couniserial dimension. We need the following lemma to give the example.

Lemma 4.11

For an ordinal number α, being of couniserial dimension α is a Morita invariant property for modules.

Proof

This is clear by the definition of couniserial dimension and [1, Proposition 21.7]. □

Example 4.12

Here we give an example of a prime right Goldie ring RR with maximal right quotient ring Qsuch that QR does not have couniserial dimension. Take R= $M_2(\mathbb{Z})$, the 2×2 matrix ring over $\mathbb{Z}$. Then R is a prime right Goldie ring with maximal right quotient ring $Q = M_2(\mathbb{Q})$. Note that under the standard Morita equivalent between the ring $\mathbb{Z}$ and $R = M_2(\mathbb{Z})$, see [17, Theorem 17.20], RR corresponds to $\mathbb{Z}\oplus\mathbb{Z}$ and so using the above lemma R has couniserial dimension 2. If $\{p_i | i \geq 1\}$ is the set of all prime numbers, then $\mathbb{Q}/\mathbb{Z} = \sum_{i=1}^{\infty} K_i/\mathbb{Z}$, where $K_i = \{m/p_i^n | n \geq 0\ m \in \mathbb{Z}\}$ and m∈Z}. Then take $Q_n = \sum_{i=n}^{\infty} K_i$. Then $M_2(Q_1) \geq M_2(Q_2) \geq \cdots$ is a descending chain of RR-submodules of Q which are not uniform R-modules. Assume that for some n, $M_2(Q_n) \cong M_2(Q_{n+1})$ with an RR-isomorphism ϕ. Let$^{\phi.\text{ Let }\phi}\begin{pmatrix}1 & 0\\ 0 & 1\end{pmatrix} =$

$\begin{pmatrix} m_1/t_1 & m_2/t_2 \\ m_3/t_3 & m_4/t_4 \end{pmatrix}$, where $m_i/t_i \in Q_{n+1}$. Suppose that $j \geq 1$ and $\phi\begin{pmatrix} 1/p_n^j & 0 \\ 0 & 1/p_n^j \end{pmatrix} =$

$\begin{pmatrix} m_{1,j}/t_{1,j} & m_{2,j}/t_{2,j} \\ m_{3,j}/t_{3,j} & m_{4,j}/t_{4,j} \end{pmatrix}$, where p_n does not odd non of $t_{i,j}$ for all $1 \leq i \leq 4$. Then since

ϕ is additive, we can easily see that $\begin{pmatrix} m_{1,j}p_n^j/t_{1,j} & m_{2,j}p_n^j/t_{2,j} \\ m_{3,j}p_n^j/t_{3,j} & m_{4,j}p_n^j/t_{4,j} \end{pmatrix} = \begin{pmatrix} m_1/t_1 & m_2/t_2 \\ m_3/t_3 & m_4/t_4 \end{pmatrix}$

ϕ is additive, we can easily see that

$$\begin{pmatrix} m_{1,j}p_n^j/t_{1,j} & m_{2,j}p_n^j/t_{2,j} \\ m_{3,i}p_n^j/t_{3,i} & m_{4,i}p_n^j/t_{4,i} \end{pmatrix} = \begin{pmatrix} m_1/t_1 & m_2/t_2 \\ m_3/t_3 & m_4/t_4 \end{pmatrix}$$

and this implies that $p_n^j | m_i$ for all $j \geq 1$ and $0 \leq i \leq 4$ and so $m_i = 0$. a contradiction. So Q_R does not have couniserial dimension.

5 SUMMARY

This paper defines couniserial dimension of a module that measures how far a module is from being uniform. The results proved in the paper demonstrate its importance for studding the structure of modules and rings and is a beginning of a larger project to study its impact. We close with some open questions:

1) Does a module with arbitrary couniserial dimension possesses indecomposable dimension?
2) Is there a theory for modules with both finite uniserial and couniserial dimensions that parallels to Krull–Schmidt–Remak–Azumaya theorem?

ACKNOWLEDGMENTS

The authors would like to thank the anonymous referee for the careful reading of this article and a lot of useful suggestions. This paper was written when the third author was visiting Ohio University, United States during May–August 2014. She wishes to express her deepest gratitude to Professor S. K. Jain for his kind guidance in her research project and Mrs. Parvesh Jain for the warm hospitality extended to her during her stay. She would also like to express her thanks to Professor E. Zelmanov for his kind invitation to visit the of University of Californian at San Diego and to give a talk.

REFERENCES

1. Anderson, F.W., Fuller, K.R.: Rings and Categories of Modules, 2nd ed. Graduate Texts in Mathematics, vol. 13. Springer, Berlin (1992)
2. Baccella, G.: On flat factor rings and fully right idempotent rings. Ann. Univ. Ferrara Sez. **26**, 125–141 (1980)
3. Baccella, G.: Von Neumann regularity of V-rings with artinian primitive factor rings. Proc. Am. Math. Soc. **103**(3), 747–749 (1988)
4. Cozzens, J., Faith, C.: Simple Noetherian Rings. Cambridge University Press, Cambridge (1975)
5. Dubrovin, N.I.: An example of a chain primitive ring with nilpotent elements. Mat. Sb. (N.S.) 120(162), no. 3, pp. 441–447 (1983) (Russian)
6. Ghorbani, A., Nazemian, Z.: On commutative rings with uniserial dimension. J. Algebra Appl. **14**(1) (2015).
7. Goldie, A.W.: Rings with Maximum Condition, Lecture Notes. Yale University, New Haven (1961)
8. Goodearl, K.R., Warfield, R.B., Jr.: An Introduction to Non commutative noetherian Rings, London Math. Soc., vol. 61. Cambridge University Press, Cambridge (2004)
9. Goodearl, K.R.: Ring Theory. Nonsingular Rings and Modules. Dekker, New York (1976)MATH
10. Goodearl, K.R.: Surjective endomorphisms of finitely generated modules. Commun. Algebra. **15**, 589–609 (1987)
11. Hart, R., Robson, J.C.: Simple rings and rings Morita equivalent to ore domains. Proc. Lond. Math. Soc. **21**(3), 232–242 (1970)
12. Gordon, R., Small, L.W.: Piecewise domains. J. Algebra **23**, 553–564 (1972)CrossRefMathSciNetMATH
13. Hirano, Y., Mogami, I.: Modules whose proper submodules are non-hopf kernels. Commun. Algebra. **15**(8), 1549–1567 (1987)
14. Hirano, Y., Poon, E., Tsutsui, H.: A generalization of complete reducibility. Commun. Algebra. **40**(6), 1901–1910 (2012)
15. Jain, S.K., Ashish Srivastava, K., Askar Tuganbaev, A.: Cyclic Modules and the Structure of Rings, Oxford Mathematical Monographs. Oxford University Press, Oxford (2012)

16. Lam, T.Y.: A crash course on stable range, cancellation, substitution, and exchange. J. Algebra Appl. **3**, 301–343 (2004)
17. Lam, T.Y.: Lectures on Modules and Rings, Graduate Texts in Mathematics, vol. 189. Springer, New York (1999)
18. McConnell, J.C., Robson, J.C.: Noncommutative Noetherian Rings. Wiley, New York (1987)
19. Michler, G.O., Villamayor, O.E.: On rings whose simple modules are injective. J. Algebra **25**, 185–201 (1973)
20. Nazemian, Z., Ghorbani, A., Behboodi, M.: Uniserial dimension of modules. J. Algebra**399**, 894–903 (2014)
21. Stoll, R.R.: Set Theory and Logic. Dover Publication Inc, New York (1961)
22. Varadarajan, K.: Anti hopfian and anti coHopfian modules. AMS Contemp. Math. Ser.**456**, 205–218 (2008)
23. Wisbauer, R.: Foundations of Module and Ring Theory. Gordon and Breach, Reading (1991)

INDEX

U

V